SpringerBriefs in Computer Science

T0236530

For further volumes:
http://www.springer.com/series/10028

Peng Li • Song Guo

Cooperative Device-to-Device Communication in Cognitive Radio Cellular Networks

 Springer

Peng Li
The University of Aizu
Aizu-Wakamatsu City
Japan

Song Guo
The University of Aizu
Aizu-Wakamatsu City
Japan

ISSN 2191-5768   ISSN 2191-5776 (electronic)
SpringerBriefs in Computer Science
ISBN 978-3-319-12594-7   ISBN 978-3-319-12595-4 (eBook)
DOI 10.1007/978-3-319-12595-4

Library of Congress Control Number: 2014956086

Springer Cham Heidelberg New York Dordrecht London

Printed on acid-free paper

Springer is part of Springer Science+Business Media (www.springer.com)

Preface

An explosion of data traffic has been observed in cellular networks in recent years due to the booming growth of mobile applications. Various mobile applications, such as video streaming, document synchronization, and social networks, on such portable devices have generated a large amount of data traffic. Device-to-device (D2D) communication has been recently proposed as a promising technique to improve resource utilization of cellular networks by offloading the traffic through base station to local direct links among devices. However, the opportunities of D2D communication are limited because of low-quality D2D links. Cooperative communication (CC) has shown great advantages in offering high capacity and reliability for wireless communication by employing several single-antenna devices to form a virtual antenna array, which motivates us to enhance the D2D communication using CC technology. The objective of this brief is to present the architecture of cooperative D2D communication, and to examine recent advances in related research topics. An extensive review of D2D communication is provided, followed by the motivations and architecture of cooperative D2D communication. We investigate the problem of maximizing the minimum transmission rate among multiple D2D pairs in cognitive radio cellular networks by jointly considering relay assignment and channel allocation. Since energy efficiency is critical for mobile devices under energy constraints, we study the lifetime maximization problem of cooperative D2D communication by an optimal dynamic allocation of resources in terms of power, channel, cooperative relay and transmission time fraction. Furthermore, we extend the proposed cooperative D2D communication by integrating new technology of network coding, and applying for a different traffic mode of broadcast. Extensive simulations results show that our proposed solutions outperforms existing work by providing significant performance enhancement to D2D communication.

August, 2014

Peng Li
Song Guo

Acknowledgements

We would like to express our gratitude to the many people who provided support, offered comments, and assisted in the editing and proofreading.

Contents

1   **Introduction**   1
     1.1   Offloading in Wireless Cellular Networks   1
     1.2   Overview of Cooperative Device-to-Device Communication   3
     1.3   Aims of This Brief   4
     References   4

2   **Literature Survey on Cooperative Device-to-Device Communication**   7
     2.1   Device-to-Device Communication   7
     2.2   Cooperative Communication   8
     References   10

3   **Cooperative Device-to-Device Communication Architecture**   13
     3.1   Cooperative Device-to-Device Communication   13
     3.2   Challenges   16
     References   17

4   **Capacity Maximization of Cooperative Device-to-Device
     Communication**   19
     4.1   Introduction   19
     4.2   System Model   21
     4.3   Problem Formulation of Joint Relay Assignment and Channel
           Allocation Problem   22
           4.3.1   Formulation of the Basic RC Problem   22
           4.3.2   Formulation of the RCNC Problem   24
           4.3.3   Hardness Analysis...................................   25
     4.4   Solution of the RC Problem   27
           4.4.1   Solving the Relaxed Problem   27
           4.4.2   Finding the Feasible Integer Solution   30
     4.5   Solution of the RCNC Problem   31
     4.6   Performance Evaluation   33
           4.6.1   Results of Example Networks   33
           4.6.2   Results of Random Networks   36

    4.7   Conclusion 38
    References 38

5 Energy Efficiency of Cooperative Device-to-Device Communication   **41**
    5.1   Introduction 41
    5.2   System Model 42
    5.3   The Problem of Max-Min Lifetime for Cooperative D2D
          Communication.. 43
    5.4   Algorithm Design 45
          5.4.1   A Basic Solution 45
          5.4.2   Reenforcement 48
    5.5   Performance Evaluation 53
          5.5.1   Simulation Setting 53
          5.5.2   Simulation Results 54
    5.6   Conclusion and Future Work 57
    References 58

6 Cooperative Device-to-Device Communication for Broadcast **61**
    6.1   Introduction 61
    6.2   System Model 62
    6.3   Relay Selection for Unicast 63
    6.4   Relay Selection for Broadcast 64
          6.4.1   Complexity Analysis 65
          6.4.2   A Greedy Relay Selection Algorithm for Broadcast 67
    6.5   Performance Evaluation 68
          6.5.1   Unicast... 69
          6.5.2   Broadcast... 71
    6.6   Conclusion 73
    References 74

7 Conclusion **77**
    7.1   Concluding Remarks.. 77
    7.2   Future Work 78

List of Abbreviations

3DM	3-dimensional matching
AF	Amplify-and-forward
CC	Cooperative communication
CRN	Cognitive radio network
D2D	Device-to-device
DF	Decode-and-forward
GRSB	Greedy relay selection algorithm for broadcast
LP	Linear programming
LTE	Long Term Evolution
MILP	Mixed-integer linear programming
MINLP	Mixed-integer nonlinear programming
MLCC	Max-min lifetime for cooperative communication
MTB-AF	Maximum throughput broadcast problem in AF relay network
MTU-AF	Maximum throughput unicast problem in AF relay network
NC	Network coding
NLP	Nonlinear programming
ORSU	Optimal relay selection algorithm for unicast
RC	Relay assignment and channel allocation
RCNC	Relay assignment and channel allocation with network coding
SNR	Signal-to-noise ratio
SPCA	Sequential parametric convex approximation

Chapter 1
Introduction

Abstract Data traffic in cellular networks has dramatically surged in recent years due to the booming growth of various mobile applications. It is hence crucial to increase network capacity to accommodate new applications and services. In this chapter, we first introduce traffic offloading technologies in wireless cellular networks. We then focus on the cooperative device-to-device communication that supports efficient traffic offloading without any infrastructure, e.g., access points, by letting a pair of devices in proximity of each other communicate over a direct link instead of through the base station. Finally, we summarize the main aims of this brief.

1.1 Offloading in Wireless Cellular Networks

Driven by the ever-increasing popularity of smartphones and tablets in recent years, wireless cellular networks have become one of the major systems accessing to the Internet with a huge number of customers due to their pervasive availability. Various mobile applications on such portable devices have generated a large amount of data traffic. For example, mobile data produced in North America was 222 PB per month in 2012 [3]. It will continually grow in the foreseeable future, as forecasted by Cisco that global mobile data traffic will increase 13-fold between 2012 and 2017 [3].

To accommodate such huge traffic demands, wireless cellular networks have been evolving to provide higher network capacity by integrating many new technologies. This evolution started from the second generation (2G) standard in early 1990s, as the first digital cellular system supporting voice services and low-speed data transmission. We are now in the fourth generation (4G) era with two candidates being actively developed today: 3GPP LTE-Advanced and IEEE 802.16m.

As an enhanced version of LTE (Long Term Evolution), LTE-Advanced [6] will meet or exceed the requirements of the 4G standard with 100 MHz bandwidth and 1Gbps peak data rate. To achieve these objectives, it integrates the emerging technologies that can be classified into two categories: one aiming to provide higher channel capacity, e.g., the multi-antennas technique and carrier aggregation, and the other focusing on developing new communication models, e.g., network traffic offloading [2].

© The Author(s) 2014
P. Li, S. Guo, *Cooperative Device-to-Device Communication
in Cognitive Radio Cellular Networks*, SpringerBriefs in Computer Science,
DOI 10.1007/978-3-319-12595-4_1

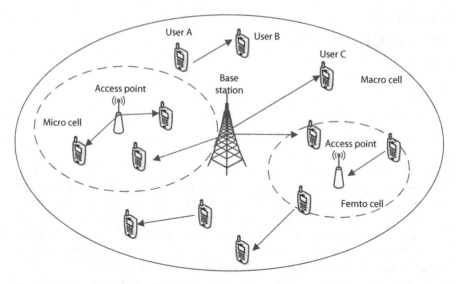

Fig. 1.1 Traffic offloading in cellular networks

Many efforts focus on improving wireless channel capacity by exploring new coding schemes or advantages of multiple antennas. For example, turbo coding/processing [1, 9] is proposed to approach the Shannon limit on channel capacity, while space-time coding [8, 10] increases the possible channel capacity by exploiting the rich multipath nature of fading wireless environments. However, they are far from solving the network capacity enhancement problem because: (1) wireless channel capacity has physical limit and cannot be increased infinitely, (2) equipping multiple antennas incurs additional hardware cost, and is not adopted by most of mobile devices, and (3) traffic growth is faster than the progress of communication technologies.

Cellular network traffic offloading provides a new communication model to the network capacity enhancement problem. It can be classified into two categories: micro/pico/femto cells, and device-to-device communication.

By deploying multiple cost-efficient access points at locations where large amounts of data are generated, cellular network capacity can be significantly increased by offloading the network traffic via the traditional cellular base stations to local low-power access points with reduced interference ranges. These access points may use different wireless technologies, such as WiFi and WiMAX, forming heterogeneous networks (HetNets) for mobile users. As shown in Fig. 1.1, traditionally, all network traffic needs to go through the base station that forms the bottleneck of the whole network. When multiple local access points are deployed, some mobile users can communicate via the local cells such that the traffic load on the macro base station can be reduced. Although traffic offloading to micro/pico/femto cells shows great advantages in increasing cellular network traffic, additional investment in network infrastructure, e.g., micro/pico/femto access points, is needed. Moreover, once these access points are deployed, they cannot be easily relocated. Since network traffic from mobile users is dynamic, i.e., traffic generation of different regions changes over time, fixed deployment would lead to low offloading performance.

As an alternative, Device-to-Device (D2D) communication [4] enables flexible traffic offloading without the support of any network infrastructure. It lets two devices in proximity of each other establish a direct local link for data transmission. After offloading some data traffic to D2D links, multiple users can transmit simultaneously in a single-collision domain. An example of D2D communication is shown in Fig. 1.1. Under cellular mode, users A and B communicate over an uplink and a downlink while all other nodes (e.g., user C) in the same cell should keep silence. When D2D communication is enabled, user A can send data to its destination B over a direct link if the link quality is good enough. Meanwhile, user C can access the network if they cause no interference to that D2D pair. The D2D communications can happen not only between a pair of nodes within a single cell, which is referred to as an intra-cell D2D pair, but also between those located in difference cells. Usually, D2D communication is managed by base stations, i.e., base stations determine whether a pair of users communicate under cellular mode or D2D mode.

1.2   Overview of Cooperative Device-to-Device Communication

The chances of D2D communication are highly dependent on the quality of D2D links. Unfortunately, without the support of a base station with powerful capability of information collection and signal processing, transmissions on D2D links are apt to be jeopardized by many factors like fading or environmental noise, and the resulting low transmission rate would reduce the incentive of adopting D2D communication for both users and network operators.

Cooperative communication (CC) [7] has shown its effectiveness in combating fading to achieve high channel capacity and reliability in a low-cost way, which is well suited to the context of D2D communication. Its basic idea is to let a relay node forward the signal received from a transmitter to its receiver such that diversity gain is obtained for the same signal traveling along different paths from direct and relay transmissions. A well-known three-node model of CC is illustrated in Fig. 1.2, where user A transmits data to user B under the assistance of a relay node. All transmissions are conducted on a frame-by-frame basis. In traditional direct transmission (DT), user A transmits data to B during the whole frame. When CC is applied, each frame is partitioned into two time slots. In the first time slot, user A transmits a signal to user B. Due to the broadcast nature of wireless communication, this transmission is also overheard by relay node. After amplifying or decoding the received signal, relay forwards it to user B in the second time slot. Finally, user B combines the received signals in two time slots from different attenuation conditions to recover the original signal.

When D2D communication is reinforced by CC, is referred to as cooperative D2D communication, new challenges, such as relay selection, channel allocation, and transmission scheduling, are raised for efficient resource allocation in cellular networks.

Fig. 1.2 A three-node model
of cooperative
device-to-device
communication

First, since a large number of D2D links would be active in a cellular network, it is
impossible to assign a dedicated channel for each of them. In a channel-constrained
network, especially when cognitive radio is applied, it is crucial to efficiently allocate
channels among all communication links, including D2D links, cellular uplinks and
downlinks, by taking CC into consideration.

Second, multiple communication links under a common channel should be carefully scheduled to guarantee a certain level of quality-of-service (QoS). This problem
becomes more difficult when CC is taken into consideration because employing a
relay node can increase link transmission rate, but would enlarge the interference
range of a link, leading to a reduced space multiplexing.

1.3 Aims of This Brief

This brief presents recent advances in cooperative D2D communication. An extensive
review of D2D communication is provided, followed by the motivations and architecture of cooperative D2D communication. We investigate the problem of maximizing
the minimum transmission rate among multiple D2D pairs in cognitive radio cellular
networks by jointly considering relay assignment and channel allocation. Since energy efficiency is critical for mobile devices under energy constraints, we study the
lifetime maximization problem of cooperative D2D communication by an optimal
dynamic allocation of resources in terms of power, channel, cooperative relay and
transmission time fraction. Furthermore, we extend the proposed cooperative D2D
communication by integrating new technology of network coding, and applying the
results to the broadcast scenario.

References

1. C. Berrou, A. Glavieux, P. Thitimajshima, Near shannon limit error-correcting coding and
   decoding: turbo-codes 1. Proc. IEEE Int. Conf. Commun. **2**, 1064–1070 (1993)
2. Y. Choi, H.W. Ji, J.yoon Park, H. chul Kim, J. Silvester, A 3w network strategy for mobile data
   traffic offloading. IEEE Commun. Mag. **49**(10), 118–123 (2011)
3. Cisco, Cisco visual networking index: global mobile data traffic forecast update, 2012–2017.
   White Paper of Cisco, http://newsroom.cisco.com (2012)
4. K. Doppler, M. Rinne, C. Wijting, C. Ribeiro, K. Hugl, Device-to-device communication as
   an underlay to lte-advanced networks. IEEE Commun. Mag. **47**(12), 42–49 (2009)

5. K. Doppler, C.-H. Yu, C. Ribeiro, P. Janis, Mode selection for device-to-device communication underlaying an lte-advanced network, in *Proceedings of IEEE Wireless Communications and Networking Conference (WCNC)*, pp. 1–6, 2010
6. A. Ghosh, R. Ratasuk, B. Mondal, N. Mangalvedhe, T. Thomas, Lte-advanced: next-generation wireless broadband technology [invited paper]. IEEE Wireless Commun. **17**(3), 10–22 (2010)
7. J. Laneman, D. Tse, G. Wornell, Cooperative diversity in wireless networks: Efficient protocols and outage behavior. IEEE Trans. Inf. Theory **50**(12), 3062–3080 (2004)
8. A. Naguib, V. Tarokh, N. Seshadri, A. Calderbank, A space-time coding modem for high-data-rate wireless communications. IEEE J. Sel. Areas Commun. **16**(8), 1459–1478 (1998)
9. E. Rosnes, O. Ytrehus, Improved algorithms for the determination of turbo-code weight distributions. IEEE Trans. Commun. **53**(1), 20–26 (2005)
10. V. Tarokh, N. Seshadri, A. Calderbank, Space-time codes for high data rate wireless communication: performance criterion and code construction. IEEE Trans. Inf. Theory **44**(2), 744–765 (1998)

Chapter 2
Literature Survey on Cooperative Device-to-Device Communication

Abstract In this chapter, we review some important work related with cooperative device-to-device communication. We first present recent advances in cellular network offloading technologies, followed by existing efforts on device-to-device communication. Then, we present the related work of cooperative communication, focusing on energy efficiency, relay assignment, and time-spectrum allocation.

2.1 Device-to-Device Communication

In the past a few years, there has been lots of research [26, 39] on network traffic offloading that focuses on offloading cellular traffic to WiFi or other networks to save bandwidth or energy. A quantitative survey of mobile data traffic surge and a strategic solution to traffic offloading has been presented in [7]. Korhonen et al. [24] have discussed existing traffic offloading solutions, and presented and evaluated three different IP traffic offloading solutions that aim to work on the internet layer and rely only on the standard IETF2 defined TCP/IP protocol suite, not requiring any access-technology-specific knowledge. Lee et al. [26] have presented a quantitative study on the performance of 3G mobile data offloading through WiFi networks. Their trace-driven simulation using the acquired whole-day traces indicates that WiFi already offloads about 65 % of the total mobile data traffic and saves 55 % of battery power without using any delayed transmission. Rstanovic et al. [39] have designed two algorithms for delay-tolerant offloading of bulky, socially recommended content from 3G networks. They find that both solutions succeed in offloading a significant amount of traffic with a positive impact on user battery lifetime.

Device-to-device (D2D) communication becomes a hot research topic recently due to its benefits of offloading data traffic at base station in cellular networks. In their early work, Janis et al. [19] propose to facilitate local peer-to-peer communication by a D2D radio that operates as an underlay network to an IMT-Advanced cellular network. Later, they have studied D2D communication in three modes, i.e., reuse mode (D2D links share common channels with cellular links), dedicated mode (D2D links use dedicated channels), and cellular mode (all communication is relayed by base station), and designed mode selection algorithm for a three-user (one D2D pair and a cellular user) cellular network [9]. Based on a similar network model, Yu et al. [51] have investigated the throughput optimization problem over shared resources

© The Author(s) 2014
P. Li, S. Guo, *Cooperative Device-to-Device Communication in Cognitive Radio Cellular Networks,* SpringerBriefs in Computer Science,
DOI 10.1007/978-3-319-12595-4_2

while fulfilling prioritized cellular service constraints. For more general models, D2D communications have been extensively investigated from aspects such as interference management, power control, spectrum sharing, and so on. For example, Janis et al. [18] have proposed a practical and efficient scheme for generating local awareness of the interference between cellular and D2D users at the base station, which then exploits the multiuser diversity inherent in the cellular network to minimize the interference. Kaufman et al. [22] have developed a distributed dynamic spectrum protocol, in which ad-hoc D2D users opportunistically access the spectrum actively in use by cellular users. A new interference management scheme is proposed to improve the reliability of D2D communication in [36]. They derive outage probability in close form and design a mode selection algorithm to minimize outage probability. Lee et al. [27] have proposed a two-stage semi-distributed resource management framework for the D2D communication. At the first stage of the framework, the base station (BS) allocates resource blocks (RBs) to BS-to-user device (B2D) links and D2D links, in a centralized manner. At the second stage, the BS schedules the transmission using the RBs allocated to B2D links, while the primary user device of each D2D link carries out link adaptation on the RBs allocated to the D2D link, in a distributed fashion. A two-tier 5G cellular network that involves a macrocell tier (i.e., BS-to-device communications) and a device tier (i.e., device-to-device communications) has been discussed in [46]. Li et al. [34] have investigated the fundamental problems of how D2D communication improves the system performance of cellular networks and what is the potential effect of D2D communication, with the aid of the optimal solutions for the system resource allocation and mode selection obtained under the realistic user and mobility conditions. Specifically, by formulating a max-flow optimization problem that maximizes the content downloading flows from all the cellular base stations to the content downloaders through any possible ways of transmission, they obtain the theoretical upper bound to system content-downloading performance.

2.2 Cooperative Communication

The basic idea of CC is proposed in the pioneering paper [47]. Later, Laneman et al. [25] have studied the mutual information and outage probability between a pair of nodes using CC under both AF (amplify-and-forward) and DF (decode-and-forward) mode. Based on their fundamental work, CC has been extensively studied from the perspectives of both physical layer and network layer. We summarize the most relevant work in the following categories: energy efficiency, relay assignment, and time-spectrum allocation in cooperative communications.

In [45], Simic et al. compare the energy-efficiency of two major cooperative diversity schemes, virtual-MISO (multiple-input-single-output) and decode-and-forward, in wireless sensor networks. They show that decode-and-forward outperforms virtual-MISO since it avoids explicit local communication among cooperating nodes. The energy efficiency of CC in wireless body area networks is investigated in [17]. To minimize the energy consumption, the problem of optimal power allocation is studied with the constraint of targeted outage probability. In [20], the energy consumption is optimized by taking amplifier power and circuit power into consideration in cooperative wireless sensor networks. An energy-efficient relay selection scheme integrated with a routing protocol is proposed in [10] for wireless sensor networks. All above work focuses on the total energy consumption of nodes involved in CC, which is significantly different from the lifetime maximization problem studied in this paper.

In [4], Bletsas et al. develop and analyze a distributed method to select the "best" relay based on local measurements of the instantaneous channel condition such that it can achieve the same diversity-multiplexing tradeoff as the protocols that require coordination and distributed space-time coding for multiple relays. Zhao et al. [53] show that it is sufficient to choose one "best" relay node instead of multiple ones for a single unicast session under AF mode. Moreover, they propose an optimal power allocation algorithm based on the best relay selection to minimize the outage probability. For multiple unicast sessions, Sharma et al. [40] consider the relay node assignment with the goal of maximizing the minimum data rate among all concurrent sessions. With the restriction that any relay node can be assigned to at most one source-destination pair, an optimal algorithm called ORA (Optimal Relay Assignment) is developed. By relaxing this constraint to allow multiple source-destination pairs to share one relay node, Yang et al. [50] prove that the total capacity maximization problem can be solved with an optimal solution within polynomial time. The benefit of CC in multi-hop wireless networks is exploited in [41] by a joint optimization of relay assignment and flow routing. When user mobility is considered, Li et al. [28] propose a dynamic relay selection scheme. With the objective of minimizing the long-term average cost while satisfying the QoS requirement. They formulate it by an optimization model based on the constrained Markov decision process and solve it using linear programming techniques.

CC in channel-constrained wireless networks is investigated in the following literatures. In [14], Gong et al. propose a cooperative relay scheme that increases the SINR at secondary receivers in cognitive radio networks. They only focus on the spectrum sharing at relay nodes under the assumption that all relay nodes are deployed at the same location. A joint optimization problem of channel pairing, channel-user assignment and power allocation is studied [15] for a dual-hop relaying network with multiple channels. It deals with a simple scenario that a source communicates with multiple users via a fixed relay. He et al. [16] optimize the resource allocation in a cognitive relay network, where a base station provides services to a set of secondary users that can assist each other using cooperative communication. A cooperative cognitive radio framework is studied in [23, 52] focusing on the interaction between the secondary and primary users. The idea is that primary users select some of secondary users to be the cooperative relays and in turn lease portion of the channel access time to them for their own data transmission. Shih et al. [44] have proposed a cooperative, multi-channel MAC protocol that incorporates the concept of cooperative communication into multi-channel MAC protocols, enabling a single transceiver to carry out the work of multiple transceivers. Recently, Li et al. [33] have studied the problem of joint relay assignment and channel allocation for cooperative communications in CRNs.

References

1. V. Asghari, S. Aissa, End-to-end performance of cooperative relaying in spectrum-sharing systems with quality of service requirements. IEEE Trans. Veh. Technol. **60**(6), 2656–2668 (2011)
2. A. Beck, A. Ben-Tal, L. Tetruashvili, A sequential parametric convex approximation method with applications to nonconvex truss topology design problems. J. Glob. Optim. **47**(1), 29–51 (2010)
3. E. Beres, R. Adve, On selection cooperation in distributed networks, in *IEEE CISS*, pp. 1056–1061, 2006
4. A. Bletsas, A. Khisti, D. Reed, A. Lippman, A simple cooperative diversity method based on network path selection. IEEE J. Sel. Area. Commun. **24**(3), 659–672 (2006)
5. D. Cabric, S. Mishra, R. Brodersen, Implementation issues in spectrum sensing for cognitive radios, in *IEEE Asilomar Conference on Signals, Systems and Computers*, pp. 772–776, 2004
6. A. Cacciapuoti, I. Akyildiz, L. Paura, Correlation-aware user selection for cooperative spectrum sensing in cognitive radio ad hoc networks. IEEE J. Sel. Area. Commun. **30**(2), 297–306 (2012)
7. Y. Choi, H.W. Ji, J. Yoon Park, H. Chul Kim, J. Silvester, A 3w network strategy for mobile data traffic offloading. IEEE Commun. Mag. **49**(10), 118–123 (2011)
8. X. Deng, A. Haimovich, Power allocation for cooperative relaying in wireless networks. IEEE Commun. Lett. **9**(11), 994–996 (2005)
9. K. Doppler, C.-H. Yu, C. Ribeiro, P. Janis, Mode selection for device-to-device communication underlaying an lte-advanced network, in *IEEE Wireless Communications and Networking Conference (WCNC)*, pp. 1–6, 2010
10. W. Fang, F. Liu, F. Yang, L. Shu, S. Nishio, Energy-efficient cooperative communication for data transmission in wireless sensor networks. IEEE Trans. Consum. Electron. **56**(4), 2185–2192 (2010)
11. M. Felegyhazi, M. Cagalj, S. Bidokhti, J.-P. Hubaux, Non-cooperative multi-radio channel allocation in wireless networks, in *Proceeding IEEE INFOCOM*, pp. 1442–1450, May 2007
12. L. Gao, X. Wang, A game approach for multi-channel allocation in multi-hop wireless networks. in *Proceeding ACM MobiHoc*, pp. 303–312, 2008
13. C. Gao, Y. Shi, Y. Hou, H. Sherali, H. Zhou, Multicast communications in multi-hop cognitive radio networks. IEEE J. Sel. Area. Commun. **29**, 784–793 (2011)
14. X. Gong, W. Yuan, W. Liu, W. Cheng, S. Wang, A cooperative relay scheme for secondary communication in cognitive radio networks, in *IEEE GLOBECOM*, pp. 1–6, 2008
15. M. Hajiaghayi, M. Dong, B. Liang, Optimal channel assignment and power allocation for dual-hop multi-channel multi-user relaying, in *IEEE INFOCOM*, pp. 76–80, April 2011
16. C. He, Z. Feng, Q. Zhang, Z. Zhang, H. Xiao, A joint relay selection, spectrum allocation and rate control scheme in relay-assisted cognitive radio system, in *IEEE VTC*, pp. 1–5, 2010
17. X. Huang, H. Shan, X. Shen, On energy efficiency of cooperative communications in wireless body area network, in *IEEE WCNC*, pp. 1097–1101, March 2011
18. P. Janis, V. Koivunen, C. Ribeiro, J. Korhonen, K. Doppler, K. Hugl, Interference-aware resource allocation for device-to-device radio underlaying cellular networks, in *Proceeding IEEE Vehicular Technology Conference*, pp. 1–5, 2009
19. P. Jänis, C.-H. Yu, K. Doppler, C. Ribeiro, C. Wijting, K. Hugl, O. Tirkkonen, V. Koivunen, Device-to-device communication underlaying cellular communications systems. Int. J. Commun. Netw. Syst. Sci. **2**(3), 169–178 (2009)
20. W. Ji, B. Zheng, Energy efficiency based cooperative communication in wireless sensor networks, in *IEEE ICCT*, pp. 938–941, Nov 2010
21. C. Jiang, Y. Shi, Y. Hou, W. Lou, Cherish every joule: maximizing throughput with an eye on network-wide energy consumption, in *Proceedings of IEEE INFOCOM*, pp. 1934–1941, March 2012
22. B. Kaufman, J. Lilleberg, B. Aazhang, Spectrum sharing scheme between cellular users and ad-hoc device-to-device users. IEEE Trans. Wirel. Commun. **12**(3), 1038–1049 (2013)

23. K. Khalil, M. Karaca, O. Ercetin, E. Ekici, Optimal scheduling in cooperate-to-join cognitive radio networks, in *IEEE INFOCOM*, pp. 3002–3010, April 2011
24. J. Korhonen, T. Savolainen, A. Ding, M. Kojo, Toward network controlled ip traffic offloading. IEEE Commun. Mag. **51**(3), 96–102 (2013)
25. J. Laneman, D. Tse, G. Wornell, Cooperative diversity in wireless networks: efficient protocols and outage behavior. IEEE Trans. Inf. Theory **50**(12), 3062–3080 (2004)
26. K. Lee, J. Lee, Y. Yi, I. Rhee, S. Chong, Mobile data offloading: how much can wifi deliver? IEEE/ACM Trans. Netw. **21**(2), 536–550 (2013)
27. D.H. Lee, K.W. Choi, W.S. Jeon, D.G. Jeong, Two-stage semi-distributed resource management for device-to-device communication in cellular networks. IEEE Trans. Wirel. Commun. **13**(4), 1908–1920, (2014)
28. Y. Li, P. Wang, D. Niyato, W. Zhuang, A dynamic relay selection scheme for mobile users in wireless relay networks, in *IEEE INFOCOM*, pp. 256–260, April 2011
29. P. Li, S. Guo, Z. Cheng, A.V. Vasilakos, Joint relay assignment and channel allocation for energy-efficient cooperative communications, in *IEEE WCNC*, pp. 5740–5744, 2012
30. P. Li, S. Guo, V. Leung, Improving throughput by fine-grained channel allocation in cooperative wireless networks, in *IEEE GLOBECOM*, pp. 5740–5744, 2012
31. P. Li, S. Guo, Y. Xiang, H. Jin, Unicast and broadcast throughput maximization in amplify-and-forward relay networks. IEEE Trans. Veh. Technol. **61**(6), 2768–2776 (2012)
32. P. Li, S. Guo, W. Zhuang, B. Ye, Capacity maximization in cooperative crns: joint relay assignment and channel allocation, in *IEEE ICC*, pp. 6619–6623, March 2012
33. P. Li, S. Guo, W. Zhuang, B. Ye, On efficient resource allocation for cognitive and cooperative communications. IEEE J. Sel. Areas Commun. **32**(2), 264–273 (2014)
34. Y. Li, Z. Wang, D. Jin, S. Chen, Optimal mobile content downloading in device-to-device communication underlaying cellular networks. IEEE Trans. Wirel. Commun. **13**(7), 3596–3608 (2014)
35. G. Liu, L. Huang, H. Xu, W. Liu, Y. Zhang, Maximal lifetime scheduling for cooperative communications in wireless networks, in *Proceeding International Conference on Computer Communications and Networks (ICCCN)*, pp. 1–6, 2010
36. H. Min, W. Seo, J. Lee, S. Park, D. Hong, Reliability improvement using receive mode selection in the device-to-device uplink period underlaying cellular networks. IEEE Trans. Wirel. Commun. **10**(2), 413–418 (2011)
37. S. Mishra, A. Sahai, R. Brodersen, Cooperative sensing among cognitive radios, in *IEEE ICC*, vol. 4, pp. 1658–1663, June 2006
38. C.H. Papadimitriou, K. Steiglitz, in *Combinatorial Optimization: Algorithms and Complexity* (Prentice-Hall, Inc., Upper Saddle River, 1982)
39. N. Ristanovic, J.-Y. Le Boudec, A. Chaintreau, V. Erramilli, Energy efficient offloading of 3g networks, in *Proceeding of IEEE Conference on Mobile Ad-Hoc and Sensor Systems (MASS)*, pp. 202–211, 2011
40. S. Sharma, Y. Shi, Y.T. Hou, S. Kompella, An optimal algorithm for relay node assignment in cooperative ad hoc networks. *IEEE/ACM Transactions on Networking*, pp. 879–892, 2010
41. S. Sharma, Y. Shi, Y.T. Hou, H.D. Sherali, S. Kompella, Cooperative communications in multi-hop wireless networks: joint flow routing and relay node assignment, in *IEEE INFOCOM*, pp. 1–9, 2010
42. S. Sharma, Y. Shi, J. Liu, Y. Hou, S. Kompella, Is network coding always good for cooperative communications? in *IEEE INFOCOM*, pp. 1–9, March 2010
43. S. Sharma, Y. Shi, Y. Hou, H. Sherali, S. Kompella, Optimizing network-coded cooperative communications via joint session grouping and relay node selection, in *IEEE INFOCOM*, pp. 1898–1906, April 2011
44. T.-C. Shih, C.-C. Kao, S.-R. Yang, A cooperative mac protocol in multi-channel wireless ad hoc networks, in *Proceeding Wireless Communications and Mobile Computing Conference (IWCMC), 2011 7th International*, pp. 1831–1836, 2011
45. L. Simic, S. Berber, K. Sowerby, Energy-efficiency of cooperative diversity techniques in wireless sensor networks, in *IEEE PIMRC*, pp. 1–5, Sept 2007

46. M. Tehrani, M. Uysal, H. Yanikomeroglu, Device-to-device communication in 5g cellular networks: challenges, solutions, and future directions. IEEE Commun. Mag. **52**(5), 86–92 (2014)
47. E.C. van der Meulen, Three-terminal communication channels. Adv. Appl. Probab. **3**, 120–154 (1971)
48. F. Wu, S. Zhong, C. Qiao, Globally optimal channel assignment for non-cooperative wireless networks, in *Proceeding IEEE INFOCOM*, pp. 1543–1551, April 2008
49. F. Wu, N. Singh, N. Vaidya, G. Chen, On adaptive-width channel allocation in non-cooperative, multi-radio wireless networks, in *Proceeding IEEE INFOCOM*, pp. 2804–2812, April 2011
50. D. Yang, X. Fang, G. Xue, Hera: an optimal relay assignment scheme for cooperative networks. IEEE J. Select. Area. Commun. **30**(2), 245–253 (2012)
51. C.-H. Yu, K. Doppler, C. Ribeiro, O. Tirkkonen, Resource sharing optimization for device-to-device communication underlaying cellular networks. IEEE Trans. Wirel. Commun. **10**(8), 2752–2763 (2011)
52. J. Zhang, Q. Zhang, Stackelberg game for utility-based cooperative cognitive radio networks, in *ACM MobiHoc*, pp. 23–32, 2009
53. Y. Zhao, R. Adve, T. Lim, Improving amplify-and-forward relay networks: optimal power allocation versus selection. IEEE Trans. Wirel. Commun. **6**(8), 3114–3123 (2007)

Chapter 3
Cooperative Device-to-Device Communication Architecture

Abstract This chapter presents the architecture of cooperative device-to-device communication in a multi-cell multi-channel wireless network. Two cooperation modes, amplify-and-forward and decode-and-forward, are introduced for cooperative communication among D2D users and relay nodes. After that, three main challenges of relay assignment, transmission scheduling and channel allocation are presented.

3.1 Cooperative Device-to-Device Communication

We consider a multi-cell multi-channel wireless network consisting of base stations (BSs), mobile stations (MSs), and relay stations (RSs) as shown in Fig. 3.1. Traditionally, all transmissions should go through base stations in a centralized communication mode. When a user communicates with the base station, others on the same channel in the same cell should stay silent to avoid interference. When D2D communication is enabled, a direct link can be established between a pair of users in proximity of each other. For example, users MS_2 and MS_3 in Fig. 3.1 can bypass the base station BS_1 and communicate directly. Compared with the traditional centralized cellular mode, multiple D2D and cellular links can be active simultaneously if they cause negligible interference to each other, leading to a significantly increased channel utilization. D2D communications happen between a pair of nodes not only within a single cell, but also in different cells. The former is referred to as an intra-cell D2D pair, e.g., $MS_2 - MS_3$, while the latter is called an inter-cell D2D pair, such as $MS_4 - MS_5$ in Fig. 3.1.

The network includes a number of relay stations, which may be deployed by network operators as dedicated ones, or contributed by third-party device providers, such as access points of local area networks or even mobile nodes that are willing to provide forwarding service. These relay stations can be employed by D2D and cellular users to improve their channel capacity. As an example shown in Fig. 3.1, relay RS_2 can assist the transmissions from node MS_2 to MS_3, forming a classical three-node model of cooperative communication. Typically, there are two cooperative communication modes, namely, amplify-and-forward (AF) and decode-and-forward (DF). The transmission processes under these two modes and direct transmission are presented as follows.

© The Author(s) 2014
P. Li, S. Guo, *Cooperative Device-to-Device Communication*
in Cognitive Radio Cellular Networks, SpringerBriefs in Computer Science,
DOI 10.1007/978-3-319-12595-4_3

13

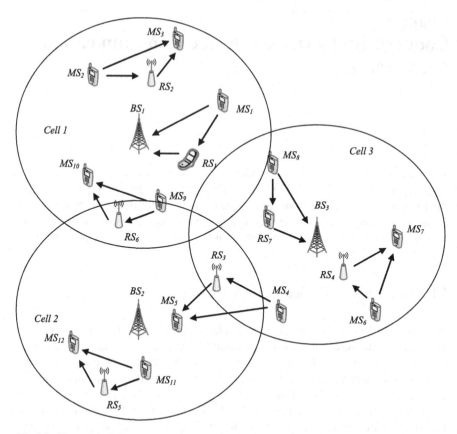

Fig. 3.1 Network infrastructure with D2D communications

Amplify-and-Forward (AF)   Under AF mode, the cooperative relay node amplifies the signal received from the source and then forwards it to the destination node. Each time frame is divided into two time slots. As shown in Fig. 3.2, in the first time slot, source s broadcasts a signal that will be received by both relay r and destination d. Then, relay r amplifies the received signal including noise to destination d in the second time slot. Although noise is amplified by cooperation, the destination receives two independently faded versions of the original signal and can make better decisions on the detection of information.

In AF mode, it is assumed that the destination knows the channel coefficients to perform optimal decoding, so some mechanisms of exchanging or estimating channel information must be incorporated into the implementation. Although there are other potential challenges, such as sampling, amplifying and retransmitting, AF is a simple method that is easy to analyze and understand the cooperative communication systems.

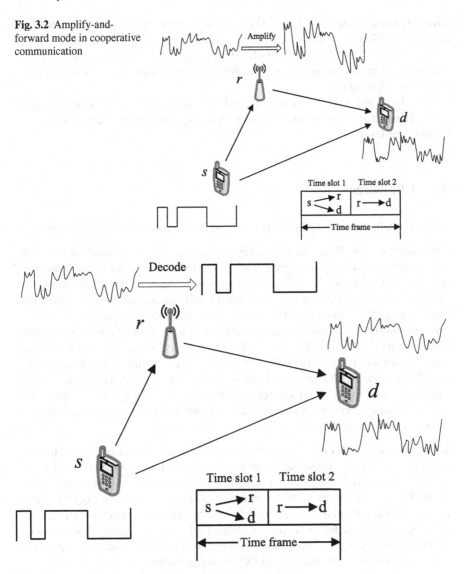

Fig. 3.2 Amplify-and-forward mode in cooperative communication

Fig. 3.3 Decode-and-forward mode in cooperative communication

Decode-and-Forward (DF) Under DF mode, the cooperative relay node decodes the received signal and re-encodes it before forwarding it to the destination node. As shown in Fig. 3.3, after receiving the signal transmitted by node s in the first time slot, relay r decodes out the original signal and then forwards it to destination. Since the destination d also receives two faded signals, it uses them to extract the original signal at the same diversity order of AF.

Direct Transmission (DT) Under DT, the transmission from source to destination takes up the whole time frame.

We observe that CC is not always better than direct transmission. In fact, a poor selection of relay node under CC could make the achievable data rate lower compared to the direct transmission.

The above physical model has resulted in several protocols in physical layer [1–7]. These protocols describe various ways through which nodes can cooperate at the physical layer.

3.2 Challenges

The main challenges in cooperative D2D communications are summarized as follows.

Relay Assignment When multiple relays are available in the network, it has been well recognized that relay assignment plays a critical role in determining the performance of CC under both AF and DF modes. Existing work [6, 5] has shown that selecting one relay is enough to achieve full diversity for a single source-destination pair. For example, Zhao et al. [5] have derived the closed-form expression of the outage probabilities when multiple relay and a single relay is employed, respectively, and shown that both probabilities are with the same order. When multiple CC sessions share a common set of relay nodes, the relay assignment should be globally optimized. For example, although RS_1 is the best relay node for both links $MS_9 - MS_{10}$ and $MS_1 - BS_1$ in Fig. 3.1, it cannot serve them simultaneously. An alternative for $MS_9 - MS_{10}$ is to choose RS_6 such that both links can achieve improved channel capacity on different channels.

Transmission Scheduling It is not practical to allocate a dedicated channel to each communication link because radio spectrum is a scarce resource that should be efficiently utilized. When multiple communication links share a common channel, they should be scheduled to guarantee a certain level of quality-of-service (QoS). When CC is applied, although channel capacity can be improved, wireless interference may become serious due to the participation of relay nodes, leading to a decreased throughput performance. Thus, the tradeoff between channel capacity improvement and wireless scheduling efficiency in CC should be studied.

Channel Allocation Channel allocation is also crucial for enhancing spectral efficiency. Under the traditional direct communication, two users can communicate when they tune to the same channel. Since a CC transmission involves three nodes, i.e., a source, a destination and a relay node, the channel allocation for cooperative D2D communication becomes more challenging as the two links use the same frequency channel in different time slots.

References

1. M. Damen, A. Hammons, Delay-tolerant distributed-tast codes for cooperative diversity. IEEE Trans. Inf. Theory **53**(10), 3755–3773 (2007)
2. D. Gunduz, E. Erkip, Opportunistic cooperation by dynamic resource allocation. IEEE Trans. Wirel. Commun. **6**(4), 1446–1454 (2007)
3. O. Gurewitz, A. de Baynast, E. Knightly, Cooperative strategies and achievable rate for tree networks with optimal spatial reuse. IEEE Trans. Inf. Theory **53**(10), 3596–3614 (2007)
4. J. Laneman, D. Tse, G. Wornell, Cooperative diversity in wireless networks: efficient protocols and outage behavior. IEEE Trans. Inf. Theory **50**(12), 3062–3080 (2004)
5. S. Savazzi, U. Spagnolini, Energy aware power allocation strategies for multihop-cooperative transmission schemes. IEEE J Sel. Area. Commun. **25**(2), 318–327 (2007)
6. A. Sendonaris, E. Erkip, B. Aazhang, User cooperation diversity. part i. system description. IEEE Trans. Commun. **51**(11), 1927–1938 (2003)
7. A. Sendonaris, E. Erkip, B. Aazhang, User cooperation diversity. part ii. implementation aspects and performance analysis. IEEE Trans. Commun. **51**(11), 1939–1948 (2003)

Chapter 4
Capacity Maximization of Cooperative Device-to-Device Communication

Abstract Cooperative communication (CC) can offer high channel capacity and reliability in an efficient and low-cost way by forming a virtual antenna array among
single-antenna nodes that cooperatively share their antennas. It has been well recognized that the selection of relay nodes plays a critical role in the performance of
multiple D2D pairs. Unfortunately, all prior work has made an unrealistic assumption that spectrum resources are unlimited and each D2D pair can communicate
over a dedicated channel with no mutual interference. In this chapter, we study the
problem of maximizing the minimum transmission rate among multiple device-to-
device communication pairs using CC in a cognitive radio network (CRN). We jointly
consider the relay assignment and channel allocation under a finite set of available
channels, where the interference must be considered. In order to improve the spectrum efficiency, we exploit the network coding opportunities existing in CC that can
further increase the capacity. Such max-min rate problems for cognitive and cooperative communications are proved to be NP-hard and the corresponding MINLP
(Mixed-Integer Nonlinear Programming) formulations are developed. Moreover, we
apply the reformulation and linearization techniques to the original optimization
problems with nonlinear and nonconvex objective functions such that our proposed
algorithms can produce high competitive solutions in a timely manner. Extensive
simulations are conducted to show that the proposed algorithms can achieve high
spectrum efficiency in terms of providing a much improved max-min transmission
rate under various network settings.

4.1 Introduction

By employing several single-antenna nodes to form a virtual antenna array, cooperative communication (CC) has been shown great advantages in offering high capacity
and reliability in wireless networks [12, 20]. Typically, there are two cooperative
communication modes, namely, amplify-and-forward (AF) and decode-and-forward
(DF). For both AF and DF, it has been well recognized that the selection of relay
nodes plays a critical role in the performance of CC. For a single device-to-device
communication pair, the full diversity order can be achieved by choosing the "best"
relay node [3, 25]. Based on this approach, an optimal relay assignment (ORA) algorithm [17] is proposed to maximize the minimum data rate among multiple D2D

© The Author(s) 2014
P. Li, S. Guo, *Cooperative Device-to-Device Communication*
in Cognitive Radio Cellular Networks, SpringerBriefs in Computer Science,
DOI 10.1007/978-3-319-12595-4_4

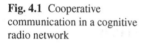

Fig. 4.1 Cooperative communication in a cognitive radio network

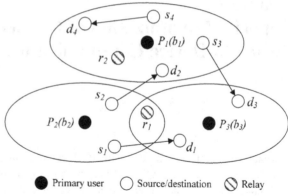

pairs. Later, a more general model [23] that allows a relay node to be shared by multiple source nodes has been studied.

Although the optimal relay assignment problem has been solved under the models in [17, 23], the spectrum efficiency has never been addressed, under an assumption that each device-to-device pair communicates over a dedicate channel without mutual interference. It is unrealistic in modern wireless networks with booming growth of various wireless applications, where the spectrum has become a scarce resource that should be efficiently utilized. Cognitive radio networks (CRNs) have been recently investigated extensively due to their potential to increase the spectrum utilization by allowing unlicensed (*i.e.*, secondary) users to opportunistically use the licensed channels as long as their transmissions do not interfere with licensed (*i.e.*, primary) users. At any time in a CRN, a set of channels that are unused by primary users can be provided for secondary users. As shown in Fig. 4.1, there are four D2D pairs and two relay nodes in a CRN with three channels b_1, b_2 and b_3, which are assigned to primary users P_1, P_2 and P_3, respectively. The transmission range of each primary user is also illustrated in the figure. Each secondary user is constrained to access a set of channels due to the activities of primary users. For example, nodes s_1 and s_2 cannot use channel b_2 since they are in the transmission range of P_2 on this channel. Obviously, existing relay assignment algorithms fail to be applied in this scenario with channel constraints. For instance, although r_1 is the best relay for s_2, it cannot assist the transmissions since they are not allowed to work on the same channel.

In this chapter, we study the problem of joint relay assignment and channel allocation (RC) for cooperative communications in CRNs. Specifically, we aim at maximizing the minimum achievable transmission rate among multiple D2D pairs with the assistance of several dedicated relay nodes. Compared with previous works [17, 23] that focus on relay assignment, this research explores the spectrum efficiency by joint optimization of channel allocation and relay assignment. Further, network coding (NC) opportunities emerge when several D2D pairs share a common relay node and thus can be applied to CC to achieve an increased spectrum efficiency. As shown in Fig. 4.1, relay r_2 can assist both s_2 and s_4 to forward signals on channel b_3. Without network coding, a frame is divided into four time slots and r_2 serves s_2 and

Fig. 4.2 Network coding for cooperation communication

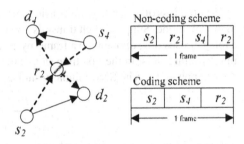

s_4 individually. When network coding is applied, only three time slots are required as shown in Fig. 4.2, in which the transmissions by s_2 and s_4 in the first two slots can be overheard by all other nodes in the same channel and then r_2 broadcasts the combined signals received from both sources in the third time slot such that d_2 and d_4 can extract their desired signals.

The main contributions are summarized as follows. First, we consider a cooperative communication model in CRNs and formulate two problems RC and RCNC (RC with Network Coding) that jointly optimize relay assignment and channel allocation with the objective of maximizing the minimum transmission rate among a give set of D2D pairs. Both problems are proved to be NP-hard. Second, we formulate the RC problem as an MINLP (Mixed-Integer Nonlinear Programming) problem and propose a low-complexity algorithm by exploiting the characteristic of the formulation. In particular, the SPCA (Sequential Parametric Convex Approximation) method [2] is applied to the relaxed problem and the results are used to find the final integer solution. Finally, extensive simulations are conducted to evaluate the proposed algorithms. The experimental results show that our proposals perform closely to the optimal solution and significantly improve the spectrum efficiency in terms of max-min transmission rate.

The rest of this chapter are organized as follows. Section 4.2 presents the system model. The joint relay assignment and channel allocation problems are formulated in Sect. 4.3. The solution for the RC and RCNC problems are elaborated in Sect. 4.4 and 4.5, respectively. Performance evaluation is given in Sect. 4.6. Finally, Sect. 4.7 concludes this chapter.

4.2 System Model

In this work, we study the data dissemination of secondary users in a CRN. Specifically, we consider a number of unicast sessions over D2D pairs (s_i, d_i), $s_i \in S = \{s_1, s_2, ..., s_n\}$ and $d_i \in D = \{d_1, d_2, ..., d_n\}$, under the support of a set of m dedicated relay nodes $R = \{r_1, r_2, ..., r_m\}$. In the following, we also use s_i to represent the unicast pair (s_i, d_i). All the nodes are equipped with a single antenna and work in a half-duplex mode that they cannot transmit and receive simultaneously.

Without loss of generality, we consider the AF mode and our results can be applied to the DF mode. Suppose each transmission between (s_i, d_i) under the assistance of a relay r_j uses time division on a frame-by-frame basis and each frame is partitioned into two time slots. In the first time slot, source s_i transmits the signal to destination d_i with power P_{s_i}. The SNR (signal-to-noise ratio) $\gamma_{s_i d_i}$ at d_i can be calculated as:

$$\gamma_{s_i d_i} = \frac{P_{s_i} |h_{s_i d_i}|^2}{\sigma_{d_i}^2}, \tag{4.1}$$

where $\sigma_{d_i}^2$ denotes the variance of background noise at d_i and $h_{s_i d_i}$ represents the effect of path-loss, shadowing and fading between s_i and d_i. Due to the broadcast nature of wireless communication, this transmission is also overheard by relay r_j. In the second time slot, relay r_j amplifies the received signal and forwards it to destination d_i. Following the analysis in [12], the mutual information I_{ij} between s_i and $d_i (1 \leq i \leq n)$ under the assistance of relay $r_j (1 \leq j \leq m)$ can be calculated by:

$$I_{ij} = \frac{1}{2} \log_2 \left(1 + \gamma_{s_i d_i} + \frac{\gamma_{s_i r_j} \gamma_{r_j d_i}}{\gamma_{s_i r_j} + \gamma_{r_j d_i} + 1} \right). \tag{4.2}$$

Under the direct transmission, source s_i transmits data to its destination in both time slots and the corresponding mutual information, denoted by I_{i0}, is:

$$I_{i0} = \log_2 \left(1 + \gamma_{s_i d_i} \right). \tag{4.3}$$

Different from most existing models, e.g., in [17, 23], where the channel resource is always sufficient, we consider a more realistic one with a finite number of available channels denoted by $B = \{b_1, b_2, ..., b_l\}$ and each channel b_k has bandwidth W_k that may be different from each other. In a CRN, each node employs some spectrum sensing techniques [5, 15] to identify a set of available channels that are not used by primary users for its communication. Due to geographical differences, the set of accessible channels at each node, denoted by $\mathcal{B}(a), a \in S \cup D \cup R$, may be different. We suppose that there is at least one common channel between s_i and d_i, i.e., $\mathcal{B}(s_i) \cap \mathcal{B}(d_i) \neq \emptyset$. Because of the channel constraint, multiple D2D pairs may work over the same channel, which shall be shared equally according to time division for the purpose of fairness [18, 23].

4.3   Problem Formulation of Joint Relay Assignment and Channel Allocation Problem

4.3.1   Formulation of the Basic RC Problem

We define a binary variable $u_{ij}(1 \leq i \leq n, 0 \leq j \leq m)$ for relay assignment as follows:

$$u_{ij} = \begin{cases} 1, & \text{if relay } r_j \text{ is assigned to pair } (s_i, d_i), \\ 0, & \text{otherwise.} \end{cases}$$

Following the discussion in [4, 25], each D2D pair is assigned at most one relay node, leading to the following constraint:

$$\sum_{j=0}^{m} u_{ij} = 1, \forall i, 1 \leq i \leq n. \tag{4.4}$$

Note that u_{i0} denotes direct transmission between s_i and d_i. To model the channel allocation, we define the following binary variables for sources and relays:

$$v_{ik} = \begin{cases} 1, & \text{if channel } b_k \text{ is allocated to pair } (s_i, d_i), \\ 0, & \text{otherwise,} \end{cases}$$

$$w_{jk} = \begin{cases} 1, & \text{if channel } b_k \text{ is allocated to relay } r_j, \\ 0, & \text{otherwise,} \end{cases}$$

where $1 \leq i \leq n$, $1 \leq j \leq m$ and $1 \leq k \leq l$. Due to the channel constraint at each node, *i.e.*, the channels occupied by primary users are not accessible, we have:

$$v_{ik} = 0, \forall b_k \in B - \mathcal{B}(s_i) \cap \mathcal{B}(d_i), \forall i, 1 \leq i \leq n, \tag{4.5}$$

$$w_{jk} = 0, \forall b_k \in B - \mathcal{B}(r_j), \forall j, 1 \leq j \leq m. \tag{4.6}$$

If CC is adopted, each D2D pair (s_i, d_i) and its associated relay r_j must be allocated a channel. Otherwise, the channel allocation at the relay node may be not necessary for direct transmission. These lead to the following constraints:

$$\sum_{k=1}^{l} v_{ik} = 1, \forall i, 1 \leq i \leq n, \tag{4.7}$$

$$\sum_{k=1}^{l} w_{jk} \leq 1, \forall j, 1 \leq j \leq m. \tag{4.8}$$

Moreover, a common channel should be assigned to the nodes in the same unicast session using either CC or direct transmission. Such a network configuration for CC has been widely adopted in the literature [1] and can be represented by:

$$u_{ij} + v_{ik} - 1 \leq w_{jk} \leq v_{ik} - u_{ij} + 1,$$
$$\forall i, j, k, 1 \leq i \leq n, 1 \leq j \leq m, 1 \leq k \leq l. \tag{4.9}$$

When relay assignment is made, *i.e.*, $u_{ij} = 1$, constraint (4.9) becomes $w_{jk} = v_{ik}(1 \leq k \leq l)$, implying that the same channel is used for s_i and r_j. Otherwise (*i.e.*, $u_{ij} = 0$), it becomes $v_{ik} - 1 \leq w_{jk} \leq v_{ik} + 1$, which is always redundant.

The transmission rate of a D2D pair (s_i, d_i) on channel b_k with the help of relay r_j can be calculated by:

$$C(s_i, b_k, r_j) \leq \frac{W_k \cdot I_{ij}}{|S(b_k)|}. \tag{4.10}$$

where $S(b_k)$ denotes the set of pairs allocated with the same channel b_k. Using our defined binary variables, we can express the transmission rate of (s_i, d_i) as:

$$C_i \leq \frac{\sum_{k=1}^{l} (v_{ik} W_k) \sum_{j=0}^{m} (u_{ij} I_{ij})}{\sum_{k=1}^{l} (v_{ik} \sum_{j=1}^{n} v_{jk})}. \tag{4.11}$$

Note that the denominator represents the number of pairs sharing a channel with (s_i, d_i). The objective of our RC problem is to find the optimal relay assignment and channel allocation that maximize the minimum capacity among all D2D pairs, i.e.,

$$\textbf{RC:} \quad \max \mathcal{C}, \quad \text{s.t.}$$

$$\mathcal{C} \leq C_i, \forall i, 1 \leq i \leq n, \tag{4.12}$$

$$(4.4), (4.5), (4.6), (4.7), (4.8), (4.9), (4.11),$$

$$u_{ij}, v_{ik}, w_{jk} \in \{0, 1\}.$$

Compared with existing works, the RC problem here is more challenging since the relay assignment and channel allocation should be jointly considered.

4.3.2 Formulation of the RCNC Problem

As in the motivation example shown in Fig. 4.2, when serving multiple D2D pairs, the relay can encode the received signals together and broadcast it to destinations by one transmission instead of forwarding the signals individually. Using network coding, the achievable transmission rate of s_i with the help of relay r_j on channel b_k can be calculated by:

$$C^{NC}(s_i, r_j, b_k) \leq \frac{W_k I_{ij}^{NC} |S(r_j)|}{|S(b_k)|(|S(r_j)| + 1)}, \tag{4.13}$$

where $S(r_j)$ denotes the set of pairs assigned a common relay node r_j. As derived in [19], the mutual information I_{ij}^{NC} when NC is applied can be calculated by:

$$I_{ij}^{NC} = \log_2 \left(1 + \gamma_{s_i d_i} + \frac{\gamma_{s_i r_j} \gamma_{r_j d_i}}{\frac{\delta_{d_i}^2}{\sigma_{d_i}^2} \sum_{k=1}^{n} u_{kj} + \gamma_{r_j d_i} + \frac{\delta_{d_i}^2}{\sigma_{d}^2} \sum_{k=1}^{n} (u_{kj} \gamma_{s_k r_j})} \right), \tag{4.14}$$

where

$$\delta_{d_i}^2 = \sigma_{d_i}^2 + \Big(\sum_{k=1}^n u_{kj} + 1 \Big) \big(\alpha_{r_j} h_{r_j d_i} \big)^2 \sigma_{r_j}^2 +$$

$$\sum_{\substack{k \in [1,n] \\ k \neq i}} \Big[u_{kj} \sigma_{d_i}^2 \big(\frac{\alpha_{r_j} h_{s_k r_j} h_{r_j d_i}}{h_{s_k d_i}} \big)^2 \Big], \tag{4.15}$$

and α_{r_j} is the amplification factor at relay node r_j [19].

We observe that the NC noise could be ignored when the background noise level is low or the relay node is shared by a small number of D2D pairs. To reduce computation complexity of (4.14), we take an approximation approach in the remaining problem formulation, in which the NC noise is ignored, leading to $I_{ij}^{NC} = 2I_{ij}$ since the content in the log operation is the same [19]. Therefore, the transmission rate of (s_i, d_i) using NC can be expressed in the optimization variables as:

$$C_i^{NC} \leq \frac{\Big(\sum_{k=1}^l (v_{ik} W_k) \Big) \Big(\sum_{j=0}^m (u_{ij} I_{ij}^{NC}) \Big) |\mathcal{S}(r_j)|}{\Big(\sum_{k=1}^l (v_{ik} \sum_{j=1}^n v_{jk}) \Big) \Big(|\mathcal{S}(r_j)| + 1 \Big)}, \tag{4.16}$$

where $|\mathcal{S}(r_j)| = \sum_{j=1}^m \big(u_{ij} \sum_{k=1}^n u_{kj} \big) + u_{i0}$ represents the number of pairs sharing the same relay node with (s_i, d_i). Eventually, the resulting formulation is:

RCNC: $\max C^{NC}$,   s.t.

$$C^{NC} \leq C_i^{NC}, \forall i, 1 \leq i \leq n, \tag{4.17}$$

$$(4.4), (4.5), (4.6), (4.7), (4.8), (4.9), (4.16),$$

$$u_{ij}, v_{ik}, w_{jk} \in \{0, 1\}.$$

The results under the DF mode can be obtained by replacing the expression of mutual information with the following formula:

$$I_{ij}^{DF} = \frac{1}{2} \min \Big\{ \log_2 (1 + \gamma_{s_i r_j}), \log_2 (1 + \gamma_{s_i d_i} + \gamma_{r_j d_i}) \Big\}.$$

That is because the proposed algorithms take the general mutual information of each D2D pair as input for both cases.

4.3.3   Hardness Analysis

We show the NP-hardness of the formulated problems by reducing the well known NP-complete 3-dimensional matching (3DM) problem to the RC problem.

Theorem 4.1   *The RC problem is NP-hard.*

Fig. 4.3 An instance of
3-dimensional matching

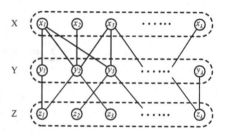

Proof In order to prove an optimization problem to be NP-hard, we need to show the NP-completeness of its decision form, which is formalized as follows.

The RC_D problem

INSTANCE: Given a set of source nodes S, a set of destination nodes D and a set of relay nodes R in a wireless network with channel set B, a constant $C \in R^+$

QUESTION: Is there a relay assignment as well as a channel allocation such that the minimum transmission rate is no less than C?

It is easy to see that the RC_D problem is in NP class as the objective function associated with a given resource allocation scheme can be evaluated in a polynomial time. The remaining proof is done by reducing the well-known 3DM problem to the RC_D problem.

The 3DM problem

INSTANCE: Given three disjoint sets X, Y and Z, where $|X| = |Y| = |Z| = \lambda$. Set $T \subseteq X \times Y \times Z$ consists of a set of 3-tuples (x_i, y_i, z_i), $x_i \in X$, $y_i \in Y$ and $z_i \in Z$.

QUESTION: Is there a subset $M \subseteq T$ such that any two 3-tuples in M are disjoint and $|M| \geq \lambda$?

For clarity, we illustrate the 3DM problem in Fig. 4.3, where nodes x_i, y_i, and z_i ($1 \leq i \leq \lambda$) represent the items in set X, Y and Z, respectively. We connect x_i and y_k as well as y_k and z_j together if $(x_i, y_j, z_k) \in T$. We now describe the reduction from 3DM to an instance of the RC_D problem. For each node $x_i \in X$, we create a D2D pair (s_i, d_i), i.e., $S = X$ and $D = X$. Each node in Y corresponds to a channel in B, i.e., $B = Y$, and all channels have the same bandwidth. The relay node set is created by letting $R = Z$. Each 3-tuple (x_i, y_j, z_k) in T specifies a configuration including a D2D pair x_i, a channel y_j, and a relay node z_k with the same transmission rate C. We also set the rate of each D2D pair under the direct transmission to be less than C. In the following, we only need to show that the 3DM problem has a solution if and only if the resulting instance of RC_D problem has a resource allocation scheme that satisfies the minimum rate requirement.

For the only-if case, we suppose that there exists a subset $M \subseteq T$ such that any two 3-tuples are disjoint and $|M| \geq \lambda$. It is a straightforward exercise to verify that the solution of the RC_D problem according to the configurations specified by M is exactly to assign each channel and relay only one D2D pair such that the capacity C of each pair can be achieved. In other words, the minimum transmission rate is no less than C.

For the if case, we suppose that the RC_D problem has a solution no less than C. In our constructed instance, the maximum rate C of each D2D pair can be achieved only when it is assigned a relay and a channel exclusively since using direct transmission and sharing channel or relay node will produce a lower transmission rate. In order to achieve the required minimum rate C, all λ D2D pairs should have the maximum rate C, which forms a solution of the 3DM problem including λ disjoint 3-tuples.

Based on the preceding analysis, we conclude that the RC_D problem is NP-complete. Thus, its optimization form RC problem is NP-hard.

For the RCNC problem, we construct an instance by setting the NC noise at a proper level such that the transmission rate under network coding is alwasy less than C. Then, the NP-hardness of RCNC problem can be proved following the similar process.

4.4 Solution of the RC Problem

Recall that the RC problem is formulated in an MINLP model. Since existing mathematical tools, such as CPLEX, do not provide a general solver for MINLP problems, we shall explore the intrinsic properties of our formulation in low-complexity algorithm design in this section. The basic idea is to relax the integer variables into continuous ones such that the global optimum solution of the resulting NLP (Nonlinear Programming) problem can be obtained. After carefully examining the formulation, we eventually convert the NLP problem into LP (Linear Programming) problem, which can be solved fast, by applying the SPCA technique [2]. If the solution of relaxed variables are integers, they are the optimal solution of the original problem as well. Otherwise, we propose a heuristic algorithm that uses the result of the relaxed problem to obtain the feasible integer solution.

4.4.1 Solving the Relaxed Problem

We observe that the formulation is in the linear form except constraint (4.11) with multiplication and division operations. Due to the fact that the logarithm function can transfer these operations into linear forms, we replace the objective function by:

$$\bar{C}_i \le \ln \left(\sum_{k=1}^{l} \left(v_{ik} W_k \right) \right) + \ln \left(\sum_{j=0}^{m} \left(u_{ij} I_{ij} \right) \right) -$$

$$\ln \left(\sum_{k=1}^{l} \left(v_{ik} \sum_{j=1}^{n} v_{jk} \right) \right), \forall i, 1 \le i \le n, \tag{4.18}$$

such that the objective function and constraint (4.12) should be changed to $\max \bar{C}$ and $\bar{C} \leq \bar{C}_i, 1 \leq i \leq n$, respectively. Note that the problem equivalency is maintained because of the monotonicity property of the logarithm function.

In the following, we consider to transfer the three nonlinear terms in (4.18) into linear forms. First of all, constraints (4.4) and (4.7) under binary variables u_{ij} and v_{ik} guarantee that the first two terms in (4.18) can be equivalently written in linear forms as:

$$\ln \left(\sum_{k=1}^{l} (v_{ik} W_k) \right) = \sum_{k=1}^{l} (v_{ik} \ln W_k), \tag{4.19}$$

$$\ln \left(\sum_{j=0}^{m} (u_{ij} I_{ij}) \right) = \sum_{j=0}^{m} (u_{ij} \ln I_{ij}). \tag{4.20}$$

To linearize the multiplication operation in the third term in (4.18), we define a new variable θ_{ik}:

$$\theta_{ik} = v_{ik} \sum_{j=1}^{n} v_{jk}, \forall i, k, 1 \leq i \leq n, 1 \leq k \leq l. \tag{4.21}$$

which represents the number of D2D pairs sharing channel b_k with s_i. Equation (4.21) can be equivalently replaced by the following linear constraints:

$$nv_{ik} - n + \sum_{j=1}^{n} v_{jk} \leq \theta_{ik} \leq \sum_{j=1}^{n} v_{jk},$$

$$\forall i, k, 1 \leq i \leq n, 1 \leq k \leq l, \tag{4.22}$$

$$0 \leq \theta_{ik} \leq nv_{ik}, \forall i, k, 1 \leq i \leq n, 1 \leq k \leq l. \tag{4.23}$$

This is because when $v_{ik} = 1$, new constraint (4.22) becomes (4.21), and (4.23) is redundant. Similarly when $v_{ik} = 0$, new constraint (4.23) becomes (4.21), and (4.22) is redundant.

Finally, we introduce a new variable η_i' to replace the third term in (4.18) and its associated constraints can be written as follows:

$$\eta_i' \leq -\ln \eta_i, \forall i, 1 \leq i \leq n, \tag{4.24}$$

$$\eta_i = \sum_{k=1}^{l} \theta_{ik}, \forall i, 1 \leq i \leq n. \tag{4.25}$$

After the above efforts on linearization, we obtain a new formulation, in which both the objective function and the constraints are expressed in linear forms except (4.24). Fortunately, after relaxing all integer variables to real number variables, the resulting problem, denoted as RRC, can be solved by an LP solvers using the SPCA

Fig. 4.4 Sequential
parametric convex
approximation for $-\ln x$
function

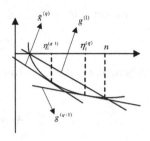

method [2], in which (4.24) is replaced by linear constraints. This conclusion is guaranteed by the property of the formulation that we developed and will be proved at the end of this subsection.

The basic idea of SPCA [2] is to iteratively solve a new LP problem by replacing the nonlinear constraints with linear ones until a converged solution (*i.e.*, the improvement is less than a given accuracy ϵ) is achieved. At each iteration, a new linear constraint is constructed such that the corresponding line is tangent to the curve defined by the nonlinear constraint at the point, which is a solution obtained in the previous iteration. The algorithm to solve the RRC problem is given in Algorithm 1.

Algorithm 1 Solve the RRC problem

1: $\mathcal{C} = -\infty$, $\mathcal{C}^{(0)} = 0$ and $q = 0$.
2: **while** $|\mathcal{C}^{(q)} - \mathcal{C}| > \epsilon$ **do**
3:    $\mathcal{C} = \mathcal{C}^{(q)}$
4:    $q = q + 1$
5:    obtain $\mathcal{C}^{(q)}$ as well as $\eta_i^{(q)}$ ($1 \leq i \leq n$) by solving the following problem with relaxed variables:

$$\textbf{LP\_RRC:} \qquad \max \bar{\mathcal{C}}, \qquad \text{s.t.}$$
$$\bar{\mathcal{C}} \leq \bar{C}_i, \forall 1 \leq i \leq n, \tag{4.26}$$
$$\bar{C}_i \leq \Big(\sum_{k=1}^{l}(v_{ik} \ln W_k)\Big) + \Big(\sum_{j=0}^{m}(u_{ij} \ln I_{ij})\Big) + \eta_i' \tag{4.27}$$
$$\text{s.t.}(4.4) - (4.9), (4.22), (4.23), (4.25), (4.29)$$

6: **end while**

In the proposed algorithm, the nonlinear constraint (4.24) is initially replaced by:

$$\eta_i' \leq \frac{-\ln n}{n-1}(\eta_i - 1), \forall i, 1 \leq i \leq n, \tag{4.28}$$

as shown by the line corssing poting $(1, 0)$ and $(n, -\ln n)$ in Fig. 4.4. The corresponding solution will serve as an upper-bound because the constraints are relaxed. Such setting guarantees to find an initial feasible solution of the RRC problem. Let $x^{(q)}$ denote the optimal solution of variable x by solving the corresponding LP problem formulated as LP_RRC at the q-th iteration of Algorithm 1. Therefore, the linear constraint at the q-th iteration can be expressed as:

$$\eta_i' \leq g^{(q)}\big(\eta_i, \eta_i^{(q-1)}\big), \forall i, 1 \leq i \leq n, \tag{4.29}$$

where function $g^{(q)}$ is given by:

$$g^{(q)}(x, x_0) = \begin{cases} \frac{-\ln n}{n-1}(x-1), & q = 1, \\ \frac{-1}{x_0}(x - x_0) - \ln x_0, & q \geq 2. \end{cases}$$

Function $g^{(q)}(q \geq 2)$ is defined by the tangent line to $-\ln x$ at point $(\eta_i^{(q-1)}, -\ln \eta_i^{(q-1)})$ as shown in Fig. 4.4.

Theorem 4.2 *The solution of RRC problem obtained by Algorithm 1 satisfies the Karush-Kuhn-Tucker (KKT) conditions.*

Proof For any feasible point $\big(\eta_i^{(q-1)}, -\ln \eta_i^{(q-1)}\big)$, we update the linear constraint $\eta_i' \leq g^{(q)}\big(\eta_i, \eta_i^{(q-1)}\big)$ for the LP_RRC formulation in the Algorithm 1. As guaranteed by the analysis in [2], the conclusion is achieved when the nonlinear function $-\ln \eta_i$ and its replaced linear function $g^{(q)}\big(\eta_i, \eta_i^{(q-1)}\big)(q \geq 2)$ have the same values at $\eta_i = \eta_i^{(q-1)}$ for both original and their first-order differential functions. These can be verified by:

$$g^{(q)}\big(\eta_i^{(q-1)}, \eta_i^{(q-1)}\big) = -\ln \eta_i^{(q-1)}, \tag{4.30}$$

$$\nabla g^{(q)}\big(\eta_i^{(q-1)}, \eta_i^{(q-1)}\big) = \nabla\big(-\ln \eta_i^{(q-1)}\big) = \frac{-1}{\eta_i^{(q-1)}}. \tag{4.31}$$

Note the KKT conditions are satisfied only for the relaxed problem, referred to as RRC in our chapter, not for the MILP problem. Although Algorithm 1 returns a solution satisfying the KKT conditions, we find out that it is always the global optimal solution empirically through extensive numerical experiments.

4.4.2 Finding the Feasible Integer Solution

If the results of variables u_{ij}, v_{ik} and w_{jk} in the solution of the RRC problem are integers, they are also the optimal solution of the original RC problem. Otherwise, they will serve as the guidance in finding the feasible integer solution. In this section, we propose a heuristic algorithm that can quickly find feasible integer solution by rounding the results returned by Algorithm 1.

Algorithm 2 Find feasible integer solution

1: **for** $i = 1$ to n **do**
2:     find $v_{ik'}$ with the largest value among v_{ik} $(1 \leq k \leq l)$
3:     set $v_{ik'} = 1, v_{ik} = 0 (1 \leq k \leq l, k \neq k')$ and $\rho(i) = k'$
4: **end for**
5: **for** i = 1 to n **do**
6:     find the u_{ij} $(1 \leq j \leq m)$ with values greater than zero and store them in set J
        according to ascending order
7:     **for** $k = 1$ to $|J|$ **do**
8:         get the $u_{ij'}$ in the k-th position in J
9:         **if** relay $r_{j'}$ can work on channel $b_{\rho(i)}$ without any conflict and improve the
            direct transmission rate of (s_i, d_i) **then**
10:            set $u_{ij'} = 1, u_{ij} = 0 (1 \leq j \leq m, j \neq j')$ and $w_{j'\rho(i)} = 1$
11:            break;
12:        **else**
13:            set $u_{ij'} = 0$
14:        **end if**
15:     **end for**
16: **end for**

The basic idea is to first make channel allocation for all unicast pairs under direct transmission, and then to assign relay node for each pair if a common channel for this CC session is available and an improved performance can be obtained. The pseudo code of the heuristic algorithm is given in Algorithm 2. In the channel allocation for each pair (s_i, d_i), we find $v_{ik'}$ with the largest value and set $v_{ik'} = 1, v_{ik} = 0(1 \leq k \leq l, k \neq k')$. Such setting is expected to achieve comparable performance to the optimal one because the real value v_{ik} would represent the probability of the corresponding channel allocation. At the same time, we save index k' of allocated channel $b_{k'}$ in $\rho(i)$. Suppose all pairs initially working under the direct transmission after channel allocation. Then, we assign relays by determining the value of each u_{ij} from line 5 to 16. For each pair (s_i, d_i), we still find the $u_{ij'}$ with the largest value among $u_{ij}(1 \leq j \leq m)$. If $r_{j'}$ can work on channel $b_{\rho(i)}$ without introducing any conflict and improve the direct transmission rate, we set $u_{ij'} = 1, u_{ij} = 0(1 \leq j \leq m, j \neq j')$ and $w_{j'b(i)} = 1$. Otherwise, we set $u_{ij'} = 0$ and continue to find another possible relay. Note that such conflict means that a relay node transmits on more than one channel simultaneously.

4.5 Solution of the RCNC Problem

In this section, we apply the similar optimization technique to solve the RCNC problem. The constraint (4.16) can be replaced by:

$$\bar{C}_i^{NC} \leq \ln \left(\sum_{k=1}^{l} \left(v_{ik} W_k \right) \right) + \ln \left(\sum_{j=0}^{m} \left(u_{ij} I_{ij} \right) \right) +$$

$$\ln \left(\sum_{j=1}^{m} \left(u_{ij} \sum_{k=1}^{n} u_{kj} \right) + u_{i0} \right) - \ln \left(\sum_{k=1}^{l} \left(v_{ik} \sum_{j=1}^{n} v_{jk} \right) \right) -$$

$$\ln\left(\sum_{j=1}^{m}\left(u_{ij}\sum_{k=1}^{n}u_{kj}\right)+u_{i0}+1\right). \qquad (4.32)$$

Comparing to the objective function of the RC problem, we notice that the additional effort is to linearize the third and fifth terms in (4.32). First of all, the multiplication operation $u_{ij}\sum_{k=1}^{n}u_{kj}$ in both term can be replaced by a new variable ϕ_{ij} as we have done to $v_{ik}\sum_{j=1}^{n}v_{jk}$ in the last section. The associated linear constraints are:

$$u_{ij}-n+\sum_{k=1}^{n}u_{kj}\leq\phi_{ij}\leq\sum_{k=1}^{n}u_{kj},$$

$$\forall i,j,1\leq i\leq n,1\leq j\leq m, \qquad (4.33)$$

$$0\leq\phi_{ij}\leq n\cdot u_{ij},\forall i,j,1\leq i\leq n,1\leq j\leq m. \qquad (4.34)$$

To linearize the third term $\ln\left(\sum_{j=1}^{m}\phi_{ij}+u_{i0}\right)$ in (4.32), we define:

$$\psi_i=\sum_{j=1}^{m}\phi_{ij}+u_{i0}, \qquad (4.35)$$

such that the non-linear constraint involved in the final formulation $\alpha_i\leq\ln\psi_i$ can be replaced by a number of linear constraints:

$$\alpha_i\leq\ln\frac{t+1}{t}(\psi_i-t)+\ln t,\forall i,t. \qquad (4.36)$$

Such constraint approximation guarantees the equivalency of the formulation because function $\ln\psi_i$ is convex and ψ_i is an integer variable.

Finally, the non-linear constraint $\beta_i\leq-\ln(\psi_i+1)$ due to the fifth term in (4.32) can be replaced by a linear constraint:

$$\beta_i\leq g^{(q)}(\psi_i+1,\psi_i^{(q-1)}+1) \qquad (4.37)$$

at the q-th iteration of the corresponding SPCA process.

To find the global optimal solution of the relaxed RCNC problem, denoted as RRCNC, the same process as in Algorithm 1 is applied except that the LP formulation LP_RRC in line 5 should be replaced by the following:

LP_RRCNC: $\max\bar{C}^{NC}$,   s.t.

$$\bar{C}^{NC}\leq\bar{C}_i^{NC},\forall i,1\leq i\leq n,(38)$$

$$\bar{C}_i^{NC}\leq\left(\sum_{k=1}^{l}\left(v_{ik}\ln W_k\right)\right)+\left(\sum_{j=0}^{m}\left(u_{ij}\ln I_{ij}\right)\right)+\alpha_i+\eta_i'+\beta_i$$

$$(4.4)-(4.9),(4.22),(4.23),(4.25),(4.29),(4.33)-(4.37)$$

To adopt to the RCNC problem, Algorithm 2 in finding the feasible integer solution is extended only in the step of relay selection in line 9 of Algorithm 2. To check the performance improvement for each (s_i,d_i), we should consider these cases: (1) direct transmission, (2) regular CC, (3) CC with network coding. The scheme with the most improvement is applied to the unicast session (s_i,d_i).

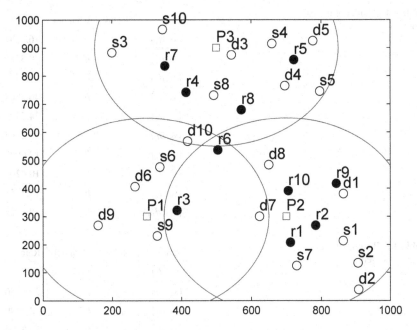

Fig. 4.5 A network with 10 D2D pairs and 10 relay nodes

4.6 Performance Evaluation

In this section, we present some numerical results to illustrate the performance of our proposed algorithms. We first study example networks to examine how an efficient resource allocation can be achieved by our proposed algorithm. Then we present the average performance over 20 random network instances each network setting with various number of n, m and l.

4.6.1 Results of Example Networks

We first consider an example network with 10 D2D pairs as well as 10 relay nodes randomly distributed within a 1000×1000 square region as shown in Fig. 4.5. Three channels b_1, b_2, and b_3 with identical bandwidth 22MHz are registered by three primary users $P1$, $P2$ and $P3$, respectively. Transmission at all source and relay nodes are made at unit power. Parameter h_{ij} describing the path-loss component between nodes i and j with a distance $||i - j||$ is given by $|h_{ij}|^2 = ||i - j||^{-4}$, in which 4 is the path-loss exponent. We set the background noise power at each node to 10^{-10} unit.

In this example network, it is easy to see that although r_6 is the best relay candidate for pair (s_8, d_8), it cannot assist the transmission since they are not allowed to work on the same channel $(\mathcal{B}(r_9) = \{b_3\}$ and $\mathcal{B}(s_8) \cap \mathcal{B}(d_8) = \{b_1\})$. The results of the

Table 4.1 Results of the RC problem

Pair	Relay	Channel	Rate
(s_1, d_1)	–	b_3	16.7251
(s_2, d_2)	–	b_3	30.8386
(s_3, d_3)	r_7	b_1	9.7373
(s_4, d_4)	–	b_2	46.3504
(s_5, d_5)	–	b_1	24.7484
(s_6, d_6)	–	b_3	29.2863
(s_7, d_7)	–	b_3	12.0030
(s_8, d_8)	r_8	b_1	10.4200
(s_9, d_9)	–	b_3	15.5257
(s_{10}, d_{10})	r_4	b_2	10.5325

Table 4.2 Results of the RCNC problem

Pair	Relay	Channel	Rate
(s_1, d_1)	–	b_3	16.7251
(s_2, d_2)	–	b_3	30.8386
(s_3, d_3)	r_4	b_2	12.9290
(s_4, d_4)	–	b_1	46.3504
(s_5, d_5)	–	b_1	24.7484
(s_6, d_6)	–	b_3	29.2863
(s_7, d_7)	–	b_3	12.0030
(s_8, d_8)	r_8	b_1	10.4200
(s_9, d_9)	–	b_3	15.5257
(s_{10}, d_{10})	r_4	b_2	14.0433

RC problem returned by our algorithm are shown in Table 4.1. Compared with the minimum transmission rate of 5.7214 under direct transmission, our algorithm can increase the transmission rate to 9.7373, by employing relay r_7 for D2D pair (s_3, d_3) under channel b_1. This pair also shares channel b_2 with pairs (s_5, d_5) and (s_8, d_8). Network coding can be applied at r_4 and our algorithm for the RCNC problem returns the minimum transmission rate of 10.4200 as shown in Table 4.2.

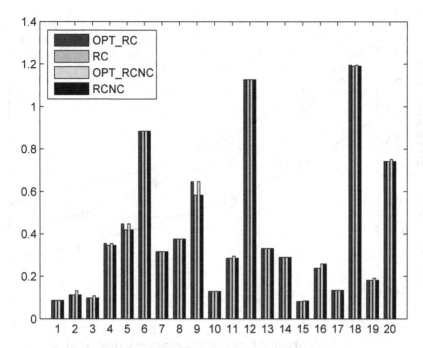

Fig. 4.6 Comparison with optimal solution in 20 random network instances

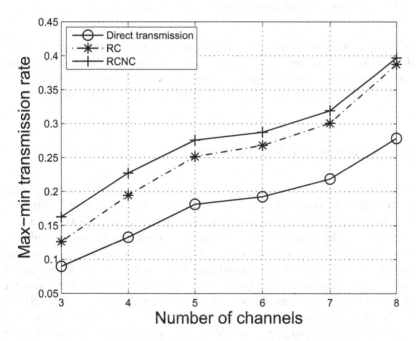

Fig. 4.7 The max-min transmission rate versus the number of channels

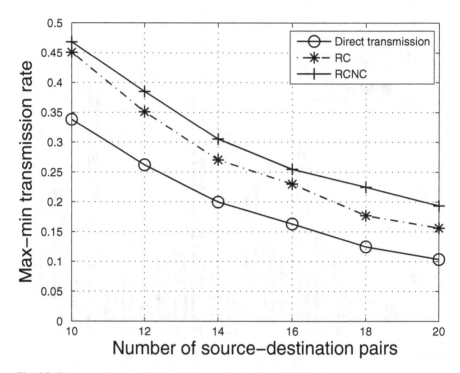

Fig. 4.8 The max-min transmission rate versus the number of D2D pairs

We also evaluate the performance of RC and RCNC by comparing their results with the optimal solutions obtained by exhaustive search in 20 random network instances that contain 8 D2D pairs, 5 relay nodes and 3 channels. As shown in Fig. 4.6, the results of our algorithms are very close to the optimal solution.

4.6.2 Results of Random Networks

We study the performance of our proposed algorithms in the form of average results over 20 random network instances. The influence of channel number is first investigated by changing its value from 3 to 8 in the networks with 15 D2D pairs and 10 relay nodes. The bandwidth of each channel is randomly distributed within the range $[20MHz, 30MHz]$ and the other simulation settings are the same as the ones used in the example study. As shown in Fig. 4.7, the performance of all the schemes increases as the channel number grows since the contention of network resources is alleviated when a larger number of channels are available. The proposed RC and RCNC always outperform the direct transmission scheme and their performance gap increases as the channel number grows. Moreover, network coding brings more gains with the smaller number of channels. The performance ratio between RC and RCNC

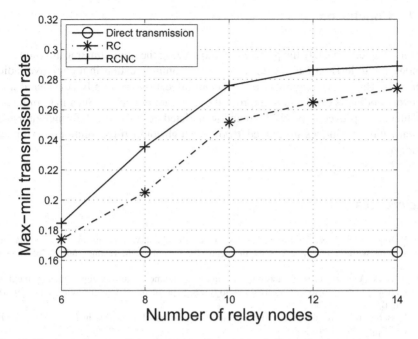

Fig. 4.9 The max-min transmission rate versus the number of relay nodes

is 1.34 in 3-channel networks. When the channel number increases to 8, this ratio decreases to 1.05. This is because more pairs may work over the same channel when the available channels are less such that the coding probability increases.

We then study the effect of the number of D2D pairs on the max-min rate. Under fixed ten relays and five available channels in the networks, as shown in Fig. 4.8, the max-min rate decreases as the pair number increases for all the transmission schemes. The performance of RC and RCNC is obviously higher than that of direct transmission and the coding gain is larger in the networks with more D2D pairs. We attribute this phenomenon to the fact that more pairs work over the same channel such that the coding probabilities increases in the networks with a larger number of D2D pairs

Finally, we evaluate the performance under different numbers of relay nodes. The number of D2D pairs and the channel number are fixed to 15 and 5, respectively. As shown in Fig. 4.9, when the number of relays is six, the RC and RCNC increase the max-min rate of the direct transmission by 5 % and 11 %, respectively. The improvement increases to 65 % and 74 %, respectively, as the number of relay nodes increases to 14. This gain comes from the assistance of a larger number of relay nodes. Moreover, we observe that increasing trends of the RC and RCNC slow down when the relay number is larger than ten. This is because each D2D pair has already found a relay node with good performance under such scenarios. Further increasing the relay number results in limited performance improvement.

4.7   Conclusion

In this chapter, we study the problem of maximizing the minimum transmission rate among multiple D2D pairs using cooperative communication in a cognitive radio network. The relay assignment and channel allocation are jointly considered and network coding is exploited to improve the spectrum efficiency. Such max-min rate problems are proved to be NP-hard and formulated as MINLPs. Reformulation and linearization techniques are applied to produce high competitive solutions in a timely manner.

References

1. V. Asghari, S. Aissa, End-to-end performance of cooperative relaying in spectrum-sharing systems with quality of service requirements. IEEE Trans. Veh. Technol. **60**(6), 2656–2668 (2011)
2. A. Beck, A. Ben-Tal, L. Tetruashvili, A sequential parametric convex approximation method with applications to nonconvex truss topology design problems. J. Global Optim. **47**(1), 29–51 (2010)
3. E. Beres, R. Adve, On selection cooperation in distributed networks, in *Proceedings IEEE CISS*, pp. 1056–1061, 2006
4. A. Bletsas, A. Khisti, D. Reed, A. Lippman, A simple cooperative diversity method based on network path selection. IEEE J. Sel. Areas Commun. **24**(3), 659–672 (2006)
5. D. Cabric, S. Mishra, R. Brodersen, Implementation issues in spectrum sensing for cognitive radios, in *Proceedings IEEE Asilomar Conference on Signals, Systems and Computers*, pp. 772–776, 2004
6. M. Felegyhazi, M. Cagalj, S. Bidokhti, J.-P. Hubaux, Non-cooperative multi-radio channel allocation in wireless networks, in *Proceedings IEEE INFOCOM*, pp. 1442–1450, May 2007
7. L. Gao, X. Wang, A game approach for multi-channel allocation in multi-hop wireless networks, in *Proceedings ACM MobiHoc*, pp. 303–312, 2008
8. X. Gong, W. Yuan, W. Liu, W. Cheng, S. Wang. A cooperative relay scheme for secondary communication in cognitive radio networks, in *Proceedings IEEE GLOBECOM*, pp. 1–6, 2008
9. M. Hajiaghayi, M. Dong, B. Liang, Optimal channel assignment and power allocation for dual-hop multi-channel multi-user relaying, in *Proceedings IEEE INFOCOM*, pp. 76–80, April 2011
10. C. He, Z. Feng, Q. Zhang, Z. Zhang, H. Xiao, A joint relay selection, spectrum allocation and rate control scheme in relay-assisted cognitive radio system, in *Proceedings IEEE VTC*, pp. 1–5, 2010
11. K. Khalil, M. Karaca, O. Ercetin, E. Ekici, Optimal scheduling in cooperate-to-join cognitive radio networks, in *Proceedings IEEE INFOCOM*, pp. 3002–3010, April 2011
12. J. Laneman, D. Tse, G. Wornell, Cooperative diversity in wireless networks: efficient protocols and outage behavior. IEEE Trans. Inf. Theory **50**(12), 3062–3080 (2004)
13. Y. Li, P. Wang, D. Niyato, W. Zhuang, A dynamic relay selection scheme for mobile users in wireless relay networks, in *Proceedings IEEE INFOCOM*, pp. 256–260, April 2011
14. P. Li, S. Guo, W. Zhuang, B. Ye, Capacity maximization in cooperative crns: joint relay assignment and channel allocation, in *Proceedings IEEE ICC*, pp. 6619–6623, June 2012
15. S. Mishra, A. Sahai, R. Brodersen, Cooperative sensing among cognitive radios, in *Proceedings IEEE ICC* Vol. 4, pp. 1658–1663, June 2006
16. S. Sharma, Y. Shi, Y.T. Hou, H.D. Sherali, S. Kompella, Cooperative communications in multi-hop wireless networks: joint flow routing and relay node assignment, in *Proceedings IEEE INFOCOM*, pp. 1–9, 2010

17. S. Sharma, Y. Shi, Y. Hou, S. Kompella, An optimal algorithm for relay node assignment in cooperative ad hoc networks. IEEE/ACM Trans. Netw. **19**(3), 879–892 (2011)
18. S. Sharma, Y. Shi, Y. Hou, H. Sherali, S. Kompella, Optimizing network-coded cooperative communications via joint session grouping and relay node selection, in *Proceedings IEEE INFOCOM*, pp. 1898–1906, April 2011
19. S. Sharma, Y. Shi, J. Liu, Y. Hou, S. Kompella, S. Midkiff. Network coding in cooperative communications: friend or foe? IEEE Trans. Mobile Comput. **11**(7), 1073–1085 (2012)
20. E.C. van der Meulen, Three-terminal communication channels. Adv. Appl. Probab. **3**, 120–154 (1971)
21. F. Wu, S. Zhong, C. Qiao, Globally optimal channel assignment for non-cooperative wireless networks, in *Proceedings IEEE INFOCOM*, pp. 1543–1551, April 2008
22. F. Wu, N. Singh, N. Vaidya, G. Chen, On adaptive-width channel allocation in non-cooperative, multi-radio wireless networks, in *Proceedings IEEE INFOCOM*, pp. 2804–2812, April 2011
23. D. Yang, X. Fang, G. Xue, Hera: An optimal relay assignment scheme for cooperative networks. IEEE J. Sel. Areas Commun. **30**(2), 245–253 2012
24. J. Zhang, Q. Zhang, Stackelberg game for utility-based cooperative cognitiveradio networks, in *Proceedings ACM MobiHoc*, pp. 23–32, 2009
25. Y. Zhao, R. Adve, T. Lim, Improving amplify-and-forward relay networks: optimal power allocation versus selection. IEEE Trans. Wirel. Commun. **6**(8), 3114–3123 (2007)

Chapter 5
Energy Efficiency of Cooperative Device-to-Device Communication

Abstract Cooperative D2D communication is able to achieve spatial diversity without requiring multiple antennas on the same node. Many efforts in exploiting the benefits of cooperative communication focus on improving the performance in terms of outage probability or channel capacity. However, the energy efficiency of cooperative D2D communication, which is critical for the mobile devices with energy constraints, has been little studied. In this chapter, we study the lifetime maximization problem for multiple D2D pairs using CC in multi-channel wireless networks by an optimal dynamic allocation of resources in terms of power, channel, cooperative relay, and transmission time fraction. We prove it NP-hard and formulate it as a mixed-integer nonlinear programming (MINLP) problem, which is then transformed into a mixed-integer linear programming (MILP) problem using linearization and reformulation techniques. By exploiting several problem-specific characteristics, a time-efficient branch-and-bound algorithm is proposed to solve the MILP problem. Extensive simulations are conducted to show that the proposed algorithm can significantly improve the performance of energy efficiency over existing solutions.

5.1 Introduction

MIMO (multiple-input and multiple-output) has shown its effectiveness in increasing network capacity by exploiting spatial diversity with multiple antennas. However, it is not always practical to equip multiple antennas on wireless devices with size and cost constraints. Cooperative communication (CC) has been proposed to achieve spatial diversity without requiring multiple antennas on the same node. By employing several single-antenna nodes to form a virtual antenna array, CC has shown great advantages in offering high capacity and reliability in wireless networks [19, 21].

Many efforts have been paid on exploiting the benefits of CC in terms of outage probability and channel capacity [7, 19, 22, 23]. For example, the outage probability and channel capacity of several CC transmission schemes are analyzed in [19]. The average signal-to-noise (SNR) and outage performance are optimized in [7]. With the objective of maximizing channel capacity, Li et al. [24] have proposed relay selection algorithms for both unicast and broadcast sessions. However, the energy efficiency of cooperative D2D communication, which is critical for the mobile devices under energy constraints, has been little studied. For instance, CC adopted by cellphones

© The Author(s) 2014

P. Li, S. Guo, *Cooperative Device-to-Device Communication*
in Cognitive Radio Cellular Networks, SpringerBriefs in Computer Science,
DOI 10.1007/978-3-319-12595-4_5

41

or laptops should provide enhanced performance in an energy-efficient manner since they are powered by batteries with limited capacity.

In addition to energy efficiency, spectrum efficiency receives increasing attention for wireless network design since spectrum has become a scarce resource under booming growth of various wireless applications. For example, there are 3 and 12 non-overlapping channels for the IEEE 802.11 b/g standards in 2.4 GHz and the IEEE 802.11a standard in 5 GHz, respectively. Moreover, some channels may be registered by licensed users, which cannot be accessed by unlicensed users if their transmissions interfere with licensed ones. A fundamental problem to improve spectrum efficiency is how to allocate the unregistered channels to unlicensed users.

In this chapter, we investigate the max-min lifetime problem for multiple energy-constrained device-to-device (D2D) pairs under the assistance of several dedicated relay nodes in multi-channel wireless networks. It is formulated as an optimization problem called MLCC (Max-min Lifetime for Cooperative Communications) with the objective of maximizing the minimum lifetime among these D2D pairs. To solve this challenging problem, we jointly consider power control, relay assignment, channel allocation, and time multiplexing for multiple D2D pairs. Our main contributions can be summarized as follows.

- First, we formulate the MLCC problem as a mixed-integer nonlinear programming (MINLP) problem and provide a theoretical analysis of its hardness.
- To provide a practical solution, we transform it into a mixed-integer linear programming (MILP) problem using linearization and reformulation techniques and then propose a branch-and-bound algorithm to solve the problem in a time-efficient manner by exploiting the problem-specific characteristics.
- Finally, extensive simulations are conducted to evaluate the performance of the proposed algorithm.

The rest of this chapter are organized as follows. Section 5.2 presents the system model. Problem formulation and hardness analysis are given in Sect. 5.3. Section 5.4 proposes an algorithm to solve the MLCC problem. Simulation results are shown in Sect. 5.5. Finally, we conclude the chapter in Sect. 5.6.

5.2 System Model

In this chapter, we consider a number of concurrent single-hop unicast sessions across a source node set $S = \{s_1, s_2, ...s_{|S|}\}$ and a destination node set $D = \{d_1, d_2, ..., d_{|D|}\}$, where $|S| = |D|$. The communication between s_l and d_l, also represented by s_l in this chapter for simplicity, must maintain a certain level of QoS (Quality-of-Service) with a transmission rate c_l. Each source $s_l \in S$ can adjust its transmission power within the range $[0, P^{max}]$ with a limited energy supply E_{s_l}. A set of dedicated relay nodes $R = \{r_1, r_2, ..., r_{|R|}\}$ with plenty of energy supplies (i.e., no energy constraints) are available in the network and they will forward data with a fixed power level P_r. A representative scenario is to apply cooperative communication to mobile devices, such as cell phones, tablets, and laptops, due to its benefits of offering high capacity and reliability in wireless networks. The adopted CC technique should work in an energy-efficient manner since mobile devices are powered by batteries with limited capacity. In addition, service providers usually deploy fixed relay nodes in the regions with high traffic to enhance network performance.

Following the discussion in [4, 40], any D2D pair is assigned to at most one relay node at a time. Each node is equipped with a single antenna and works in a half-duplex mode, i.e., it cannot transmit and receive simultaneously. Similar to the assumption made in [11, 24], the channel response is considered independent in this chapter. It is define as $\beta_{xy} = \frac{|h_{xy}|^2}{\sigma_y^2}$, where σ_y^2 denotes the received background noise power at node y, and h_{xy} denotes the channel coefficient representing the effect of path-loss, shadowing and fading between nodes x and y. Following the analysis in [19], the mutual information between s_l and d_l with the assistance of relay $r_m \in R$ under DF mode is calculated by:

$$I(s_l, r_m) = \frac{1}{2}\min\left\{\log_2\left(1 + \beta_{s_l r_m} P_{s_l}\right),\right.$$
$$\left.\log_2\left(1 + \beta_{s_l d_l} P_{s_l} + \beta_{r_m d_l} P_r\right)\right\}. \tag{5.1}$$

As a special case, the mutual information between source s_l and its destination d_l under DT is:

$$I(s_l, \emptyset) = \log_2\left(1 + \beta_{s_l d_l} P_{s_l}\right). \tag{5.2}$$

A set of channels $B = \{b_1, b_2, ...b_{|B|}\}$ with identical bandwidth W are available in the network and each node $x(x \in S \cup D \cup R)$ can only access a subset denoted by $\mathcal{B}(x)$, which may differ depending on its geographical position. Due to the limited channel resource in the network, multiple sessions would work on the same channel. We consider a single-collision-domain based channel model, in which all transmissions will interfere with each other if they are under the same channel. This model has been widely accepted and used in the recent literature for related topic in cooperative communications [32, 38] and spectrum efficiency [9, 10, 36, 37]. Under this model, multiple D2D pairs would share a channel using time-division multiplexing in a periodic transmission pattern, i.e., a superframe, with T CC/DT frames.

5.3 The Problem of Max-Min Lifetime for Cooperative D2D Communication

Given a network instance, let Φ be the set with all possible S-R pairs. Each S-R pair $\phi_i \in \Phi$ consists of a source, a destination and a relay, denoted as $s(\phi_i)$, $d(\phi_i)$ and $r(\phi_i)$, respectively, working under the same channel, i.e., $\mathcal{B}(s(\phi_i)) \cap \mathcal{B}(d(\phi_i)) \cap \mathcal{B}(r(\phi_i)) \neq \emptyset$. In the MLCC problem, the network resource in both frequency and time domains can be represented by a matrix as shown in Fig. 5.1. To describe the network resource allocation for S-R pairs, we define a binary variable u_{ij}^k as follows:

Fig. 5.1 Network resource

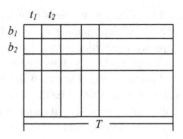

$$u_{ij}^k = \begin{cases} 1, & \text{if } \phi_i \text{ works in the } j\text{-th frame} \\ & \text{under channel } b_k, \\ 0, & \text{otherwise.} \end{cases}$$

For any feasible allocation scheme, the following constraints must hold.

$$\sum_{\phi_i \in \Phi} u_{ij}^k \le 1, \forall b_k \in B, 1 \le j \le T \tag{5.3}$$

$$\sum_{s(\phi_i)=s_l} \sum_{b_k \in B} u_{ij}^k \le 1, \forall s_l \in S, 1 \le j \le T \tag{5.4}$$

$$\sum_{r(\phi_i)=r_m} \sum_{b_k \in B} u_{ij}^k \le 1, \forall r_m \in R, 1 \le j \le T \tag{5.5}$$

Constraint (5.3) represents the fact that each frame under any channel can accommodate at most one S-R pair. Because of the single-antenna setting, each source and relay node in the network cannot work on multiple channels simultaneously as shown in constraints (5.4) and (5.5), respectively.

Let P_{ij}^k be the transmission power of source node $s(\phi_i)$ scheduled in frame j under channel b_k. The relationship between P_{ij}^k and u_{ij}^k can be represented by:

$$0 \le P_{ij}^k \le u_{ij}^k P^{max}, \forall \phi_i \in \Phi, \forall b_k \in B, 1 \le j \le T. \tag{5.6}$$

The mutual information I_{ij}^k under transmission power P_{ij}^k is constrained by either

$$I_{ij}^k \le 0.5 \log_2 \left(1 + \beta_{s(\phi_k)r(\phi_k)} P_{ij}^k \right),$$
$$\forall \phi_k \in \Phi, r(\phi_i) \ne \emptyset, \forall b_k \in B, 1 \le j \le T, \tag{5.7}$$
$$I_{ij}^k \le 0.5 \log_2 \left(1 + \beta_{r(\phi_k)d(\phi_k)} P_r + \beta_{s(\phi_k)d(\phi_k)} P_{ij}^k \right),$$
$$\forall \phi_k \in \Phi, r(\phi_i) \ne \emptyset, \forall b_k \in B, 1 \le j \le T \tag{5.8}$$

under CC because of (5.1), or

$$I_{ij}^k \le \log_2 \left(1 + \beta_{s(\phi_k)d(\phi_k)} P_{ij}^k \right),$$

$$\forall \phi_k \in \Phi, r(\phi_i) = \emptyset, \forall b_k \in B, 1 \le j \le T \qquad (5.9)$$

under DT because of (5.2).

Finally, the network resource assigned to each pair (s_l, d_l) must guarantee the transmission rate c_l to be achieved, i.e.,

$$c_l \le \frac{W}{T} \sum_{j=1}^{T} \sum_{s(\phi_i)=s_l} \sum_{b_k \in B} I_{ij}^k, \forall s_l \in S. \qquad (5.10)$$

The objective of MLCC problem is to find the optimal dynamic power control, relay assignment, and channel allocation over the superframe such that the minimum lifetime of all $|S|$ concurrent D2D pairs is maximized. By defining a variable L to represent the minimum lifetime among all D2D pairs, the MLCC problem can be formally presented as follows.

MLCC   max L, subject to:

$$L \le \frac{TE_{s_l}}{\sum_{j=1}^{T} \sum_{s(\phi_i)=s_l} \sum_{b_k \in B} P_{ij}^k}, \forall s_l \in S, \qquad (5.11)$$

$$(5.3) - (5.10).$$

We observe that above formulation is a MINLP problem due to the logarithm functions in constraints (5.7–5.9). This problem cannot be efficiently solved because it combines the difficulties from nonlinear constraints and integer variables.

Theorem 5.1 *The MLCC problem is NP-hard.*

Proof The MLCC problem is shown to be NP-hard by reducing the well-known 3-dimensional matching (3DM) problem [28] following the similar proof of Theorem 4.1

5.4 Algorithm Design

5.4.1 A Basic Solution

The basic idea of solving the MLCC problem is to first transfer the MINLP formulation into a mixed integer linear programming (MILP) using linearization techniques. Then, we propose a time-efficient algorithm to solve it based on a branch-and-bound framework.

We first consider to linearize constraint (5.7) that can be rewritten in the following form:

$$\frac{e^{(2\ln 2)I_{ij}^k} - 1}{\beta_{s(\phi_i)r(\phi_i)}} \le P_{ij}^k,$$

$$\forall \phi_i \in \Phi, r(\phi_i) \ne \emptyset, \forall b_k \in B, 1 \le j \le T. \qquad (5.12)$$

Fig. 5.2 Approximation of an
exponential function

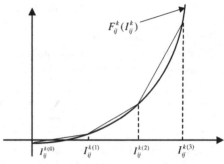

We denote the left-hand side of above constraint as $F_{ij}^k\big(I_{ij}^k\big)$. To linearize constraint (5.12), we use a set of Q line segments $f_{ij}^{k(q)}\big(I_{ij}^k\big)$, $1 \leq q \leq Q$, to approximate $F_{ij}^k\big(I_{ij}^k\big)$ as shown in Fig. 5.2, where the line segments are determined by evenly distributed points $I_{ij}^{k(q)}$. The q-th segment can be formally expressed as:

$$f_{ij}^{k(q)}\big(I_{ij}^k\big) = \alpha_{ij}^{k(q)}\big(I_{ij}^k - I_{ij}^{k(q-1)}\big) + F_{ij}^{k(q-1)}, \tag{5.13}$$

where $\alpha_{ij}^{k(q)}$ is the slope of the q-th segment and can be calculated by:

$$\alpha_{ij}^{k(q)} = \frac{F_{ij}^{k(q)} - F_{ij}^{k(q-1)}}{I_{ij}^{k(q)} - I_{ij}^{k(q-1)}}. \tag{5.14}$$

After such an approximation process, nonlinear constraint (5.12) can be replaced by a set of linear ones:

$$f_{ij}^{k(q)}\big(I_{ij}^k\big) \leq P_{ij}^k, \forall \phi_i \in \Phi, r(\phi_i) \neq \emptyset,$$

$$\forall b_k \in B, 1 \leq j \leq T, 1 \leq q \leq Q. \tag{5.15}$$

We rewritten constraints (5.8) and (5.9) into similar forms of (5.12) and denote their left-hand sides as $G_{ij}^k(I_{ij}^k)$ and $H_{ij}^k(I_{ij}^k)$, respectively. By applying the same approach, $G_{ij}^k(I_{ij}^k)$ and $H_{ij}^k(I_{ij}^k)$ can be approximated by a set of segments $g_{ij}^{k(q)}$ and $h_{ij}^{k(q)}$, $1 \leq q \leq Q$, respectively. The linearized constraints for (5.8) and (5.9) are therefore:

$$g_{ij}^{k(q)}\big(I_{ij}^k\big) \leq P_{ij}^k, \forall \phi_i \in \Phi, r\big(\phi_i\big) \neq \emptyset,$$

$$\forall b_k \in B, 1 \leq j \leq T, 1 \leq q \leq Q, \tag{5.16}$$

$$h_{ij}^{k(q)}\big(I_{ij}^k\big) \leq P_{ij}^k, \forall \phi_i \in \Phi, r\big(\phi_i\big) \neq \emptyset,$$

$$\forall b_k \in B, 1 \leq j \leq T, 1 \leq q \leq Q, \tag{5.17}$$

respectively.

By setting a new objective L' as the reciprocal of L, we finally obtain an MILP for the MLCC problem as follows.

Algorithm 3 Solving the MLCC_L problem

1: $\mathcal{P} = \{p_0\}, \mathcal{U} = \infty$;
2: set $\bar{L}_{p_0}$ as the optimal solution of the relaxed problem p_0;
3: **while** $\mathcal{P} \neq \emptyset$ **do**
4:      select a problem $p \in \mathcal{P}$ with the minimum $\bar{L}_p$ and let $\mathcal{L} = \bar{L}_p$;
5:      set U_p as the solution of p by rounding;
6:      **if** $U_p < \mathcal{U}$ **then**
7:          $U^* = U_p, \mathcal{U} = U_p$;
8:          **if** $\mathcal{L} \geq (1 - \epsilon)\mathcal{U}$ **then**
9:              **return** the $(1 - \epsilon)$-optimal solution U^*;
10:          **else**
11:              remove all problems $p' \in \mathcal{P}$ with $\bar{L}_{p'} \geq (1 - \epsilon)\mathcal{U}$;
12:          **end if**
13:      **end if**
14:      construct two problems p_1 and p_2 according to formulation MLCC_L;
15:      Find an unfixed u_{ij}^k with maximum value in the result of p;
16:      In problem p_1, set $u_{ij}^k = 1$ and fix associated variables according to (5.3) - (5.5);
17:      In problem p_2, set $u_{ij}^k = 0$ and fix associated variables according to (5.3) - (5.5);
18:      solve problems p_1 and p_2 by relaxing all unfixed integer variables and obtain the results $\bar{L}_{p_1}$ and $\bar{L}_{p_2}$;
19:      **if** $\bar{L}_{p_1} < (1 - \epsilon)\mathcal{U}$, **then** put p_1 into $\mathcal{P}$; **end if**
20:      **if** $\bar{L}_{p_2} < (1 - \epsilon)\mathcal{U}$, **then** put p_2 into $\mathcal{P}$; **end if**
21: **end while**
22: **return** the $(1 - \epsilon)$-optimal solution U^*;

$$\textbf{MLCC\_L} \qquad\qquad \min L', \text{ subject to:}$$

$$L' \geq \frac{\sum_{j=1}^{T} \sum_{s(\phi_i) = s_l} \sum_{b_k \in B} P_{ij}^k}{TE_{s_l}}, \forall s_l \in S, \tag{5.18}$$

$$(5.3) - (5.6), (5.10), (5.15), (5.16), \text{ and } (5.17).$$

We observe that the larger the value of Q is set to, the smaller the approximation error is produced. However, it will lead to more constraints that increase the complexity.

To deal with the integer variables in MLCC_L, we propose an algorithm based on a branch-and-bound framework as formally described in Algorithm 3, in which $\mathcal{P}$ represents a problem set with a lower bound $\mathcal{L}$ and an upper bound $\mathcal{U}$ of the optimal solution. Initially, $\mathcal{P}$ only includes the original problem denoted by p_0 that is constructed by relaxing all the integer variables in MLCC_L formulation. For any problem $p \in \mathcal{P}$, the optimal solution of the corresponding relaxed problem can be obtained by solving the linear programming and it can serve as a lower bound, denoted as $\bar{L}_p$, of the solution to the original problem. Then, the algorithm proceeds iteratively as follows. In each round, we find a problem $p \in \mathcal{P}$ with minimum $\bar{L}_p$ and then set $\mathcal{L} = \bar{L}_p$. While any feasible solution of p can serve as an upper bound, the one obtained by rounding under the satisfaction of all constraints is used and denoted by U_p. The tightest upper bound $\mathcal{U}$ is updated from line 6 to 13. If the performance gap between $\mathcal{L}$ and $\mathcal{U}$ is less than a predefined small number ϵ, a $(1 - \epsilon)$-optimal solution U^* is returned. Otherwise, we replace problem p with two subproblems p_1 and p_2 constructed by branching binary variables $u_{ij}^k, \phi_i \in \Phi, b_k \in B, 1 \leq j \leq T$. Other variables can be quickly determined after they have been fixed because the resulting problem becomes a tractable linear programming (LP) problem. In particular, we fix integer variables u_{ij}^k in a decreasing order of their solutions to the relaxed problem p, which is a greedy approach similar to rounding such that the solution with desired performance could be reached as early as possible. By fixing one variable, other integer variables could be fixed right away as many as possible due to the constraints regarding these correlated variables. Specifically, after a variable u_{ij}^k is selected to branch, we set its value to 1 and 0 in subproblem p_1 and p_2, respectively, as shown in lines 16 and 17. Simultaneously, other variables are fixed according to constraints (5.3–5.5). Finally, we put subproblems p_1 and p_2 into the problem set $\mathcal{P}$ if their solutions are less than $(1 - \epsilon)\mathcal{U}$, as shown in lines 19 and 20.

5.4.2   Reenforcement

In this subsection, we exploit some problem-specific characteristics to improve the execution efficiency of the basic solution proposed in last subsection. By carefully examining MLCC_L, we find that its high complexity stems from constraints (5.7–5.9) in the original MLCC, at an order of $O(|\Phi| \cdot |B| \cdot T)$, and the linear-constraint approximation further increases this order by Q times. In the following subsections, we reduce the complexity of these two parts respectively.

Simplified Formulation

We notice that all channels are equivalent based on our network model. It motivates us to simplify the MLCC formulation by defining a variable P_i for power control of any S-R pair ϕ_i such that variables P_{ij}^k can be replaced by $u_{ij}^k P_i$, i.e.,

$$P_{ij}^k = u_{ij}^k P_i, \forall \phi_i \in \Phi, \forall b_k \in B, 1 \leq j \leq T. \tag{5.19}$$

Similarly, we define $I_{ij}^k = u_{ij}^k I_i$. Therefore, constraints (5.7–5.9) can be simplified as:

$$I_i \leq 0.5 \log_2 \left(1 + \beta_{s(\phi_k)r(\phi_k)} P_i\right),$$
$$\forall \phi_k \in \Phi, r(\phi_i) \neq \emptyset, \tag{5.20}$$
$$I_i \leq 0.5 \log_2 \left(1 + \beta_{r(\phi_k)d(\phi_k)} P_r + \beta_{s(\phi_k)d(\phi_k)} P_i\right),$$

$$\forall \phi_k \in \Phi, r(\phi_i) \neq \emptyset, \tag{5.21}$$

$$I_i \leq \log_2 \left(1 + \beta_{s(\phi_k)d(\phi_k)} P_i\right), \forall \phi_k \in \Phi, r(\phi_i) = \emptyset, \tag{5.22}$$

with a reduced complexity at order of only $O(|\Phi|)$.

The introduced nonlinear constraint (5.19) can be replaced by:

$$0 \leq P_{ij}^k \leq u_{ij}^k P^{max}, \forall \phi_i \in \Phi, \forall b_k \in B, 1 \leq j \leq T, \tag{5.23}$$

$$P_i - P^{max}\left(1 - u_{ij}^k\right) \leq P_{ij}^k \leq P_i,$$

$$\forall \phi_i \in \Phi, \forall b_k \in B, 1 \leq j \leq T. \tag{5.24}$$

The equivalence holds for the following reasons. When $u_{ij}^k = 1$, both constraints (5.19) and (5.24) become $P_{ij}^k = P_i$, and (5.23) is redundant. When $u_{ij} = 0$, both constraints (5.19) and (5.23) become $P_{ij}^k = 0$, and (5.24) is redundant. In a similar way, $I_{ij}^k = u_{ij}^k I_i$ is linearized as:

$$0 \leq I_{ij}^k \leq I_i^{max} u_{ij}^k, \forall \phi_i \in \Phi, \forall b_k \in B, 1 \leq j \leq T, \tag{5.25}$$

$$I_i - I_i^{max}\left(1 - u_{ij}^k\right) \leq I_{ij}^k \leq I_i,$$

$$\forall \phi_i \in \Phi, \forall b_k \in B, 1 \leq j \leq T, \tag{5.26}$$

where I_i^{max} is the maximum mutual information of ϕ_i using power P^{max}. Finally, we obtain a new formulation **MLCC_S** written as follows:

MLCC_S   min L', subject to:

$$(5.3) - (5.6), (5.10), (5.18), \text{and } (5.20) - (5.26).$$

Simplified Linear Approximation

Following a similar linearization technique in subsection 5.1, the nonlinear constraints (5.20–5.22) in above MLCC_S formulation can be approximated by a set of line segments that evenly distributed within $[0, I_i^{max}]$. However, this straightforward approach would generate too many constraints to guarantee a certain approximation error. To reduce the approximation complexity, we propose an algorithm, which is similar to the piece-wise linear approximation approach for logarithm function in [17], to find the minimum Q such that the maximum approximation error of each line segment is bounded by ξ. The design of this algorithm is based on the observation that an exponential function can be approximated in the same error bound by using less line segments when I_i is small but more when I_i is large because the derivation of the exponential function is an increasing function of I_i.

Algorithm 4 Approximating exponential function

1: $q = 0$;
2: $I_i^{(q)} = 0$;
3: **for** $I_i^{(q)} \leq I_i^{max}$ **do**
4:    $q = q + 1$;
5:    find $I_i^{(q)}$ that satisfies:

$$\alpha_i^{(q)}(\mathcal{I}_i^{(q)} - I_i^{(q-1)}) + \bar{F}_i(I_i^{(q-1)}) - \bar{F}_i(I_i^{(q)}) = \xi,$$

   where $\mathcal{I}_i^{(q)} = \frac{1}{2}\log_2 \frac{\beta_{s(\phi_i)r(\phi_i)}\alpha_i^{(q)}}{2\ln 2}$.
6:    create a line $\bar{f}_i^{(q)}(I_i) = \alpha_i^{(q)}(I_i - I_i^{(q-1)}) + \bar{F}_i(I_i^{(q-1)})$
7: **end for**

Specifically, we first consider constraint (5.20) that can be rewritten as:

$$\frac{e^{(2\ln 2)I_i} - 1}{\beta_{s(\phi_i)r(\phi_i)}} \leq P_i, \forall \phi_i \in \Phi, r(\phi_i) \neq \emptyset, \tag{5.27}$$

whose left-hand side is denoted by $\bar{F}_i$. The formal description of approximation algorithm is shown in Algorithm 4.

In Algorithm 4, we start from the first point $I_i^{(0)} = 0$ and find a set of line segments in **for** loop from line 3 to 7. In each iteration, given point $I_i^{(q-1)}$ that is not greater that I_i^{max}, we find the next point $I_i^{(q)}$ according to the rule in line 5, where $\alpha_i^{(q)}$ denotes the slope of the line segment between $I_i^{(q-1)}$ and $I_i^{(q)}$:

$$\alpha_i^{(q)} = \frac{\bar{F}_i\big(I_i^{(q)}\big) - \bar{F}_i\big(I_i^{(q-1)}\big)}{I_i^{(q)} - I_i^{(q-1)}}. \tag{5.28}$$

After obtaining the value of $I_i^{(q)}$, we create a line segment $\bar{f}_i^{(q)}(I_i)$ as shown in line 6. Finally, we obtain a set of line segments $\bar{f}_i^{(q)}(I_i)$ that approximate $\bar{F}_i(I_i)$ with an error bound ξ. This is guaranteed by Theorem 2.

Theorem 5.2 *The maximum approximation error of each linear segment returned by Algorithm 4 is at most ξ.*

Proof Consider the q-th segment $\bar{f}_i^{(q)}(I_i)$ as shown in Fig. 5.3, the maximum approximation error is achieved at the tangential point of a line that is parallel with $\bar{f}_i^{(q)}(I_i)$. To obtain the tangential point, denoted by $\mathcal{I}_i^{(q)}$, we let the derivative of the exponential function be equal to the slope of $\bar{f}_i^{(q)}(I_i)$, i.e.,

$$\left(\frac{e^{(2\ln 2)I_i} - 1}{\beta_{s(\phi_i)r(\phi_i)}}\right)' = \frac{(2\ln 2)2^{2I_i}}{\beta_{s(\phi_i)r(\phi_i)}} = \alpha_i^{(q)}. \tag{5.29}$$

Fig. 5.3 Maximum
approximation error

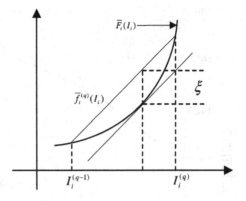

Because $\mathcal{I}_i^{(q)}$ is a solution of the above equation, after solving it, we have

$$\mathcal{I}_i^{(q)} = \frac{1}{2}\log_2 \frac{\beta_{s(\phi_i)r(\phi_i)}\alpha_i^{(q)}}{2\ln 2}. \tag{5.30}$$

Thus, the maximum approximation error can be calculated by:

$$\xi = \bar{f}_i^{(q)}\left(\mathcal{I}_i^{(q)}\right) - \bar{F}_i\left(\mathcal{I}_i^{(q)}\right)$$

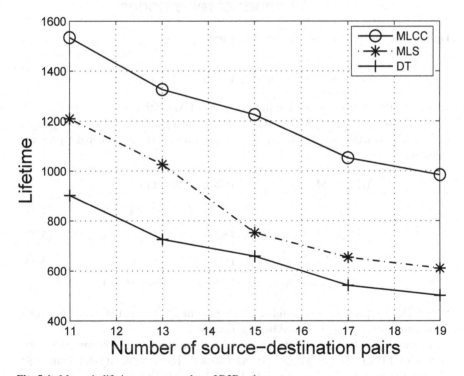

Fig. 5.4 Max-min lifetime versus number of D2D pairs

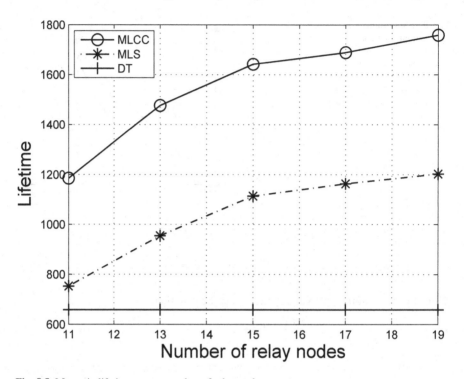

Fig. 5.5 Max-min lifetime versus number of relay nodes

$$= \alpha_i^{(q)}\left(\mathcal{I}_i^{(q)} - I_i^{(q-1)}\right) + \bar{F}_i\left(I_i^{(q-1)}\right) - \bar{F}_i\left(\mathcal{I}_i^{(q)}\right),$$

which exactly is the condition required in line 5 of Algorithm 4.

Constraints (5.21) and (5.22) can be linearized in a similar way by line segments $\bar{g}_i^{(q)}$ and $\bar{h}_i^{(q)}$ such that the MLCC_S problem can be transferred into an MILP problem as follows

MLCC_SL   min L', subject to:

$$\bar{\tilde{f}}_i^{(q)} \leq P_i, \forall \phi_k \in \Phi, r(\phi_i) \neq \emptyset, 1 \leq q \leq Q(\bar{F}_i), \qquad (5.31)$$

$$\bar{g}_i^{(q)} \leq P_i, \forall \phi_k \in \Phi, r(\phi_i) \neq \emptyset, 1 \leq q \leq Q(\bar{G}_i), \qquad (5.32)$$

$$\bar{h}_i^{(q)} \leq P_i, \forall \phi_k \in \Phi, r(\phi_i) = \emptyset, 1 \leq q \leq Q(\bar{H}_i), \qquad (5.33)$$

$$(5.3) - (5.6), (5.10), (5.18), \text{ and } (5.23) - (5.26),$$

where $Q(\cdot)$ represents the number of line segments to approximate the input exponential function $(\cdot)$ obtained by Algorithm 4.

To show the efficiency of the proposed reformulation, we compare the complexity of MLCC_L and MLCC_SL as follows. The complexity-dominant part (5.15–5.17) in MLCC_L becomes (5.31–5.33) in MLCC_SL with a significant

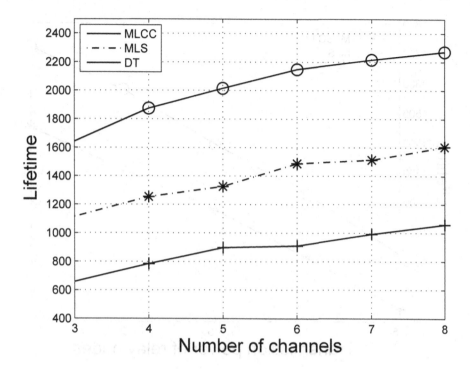

Fig. 5.6 Max-min lifetime versus number of channels

simplification from $O(|\Phi| \cdot |B| \cdot T \cdot Q)$ down to $O(|\Phi| \cdot Q')$, where $Q' = \max_{i=1}^{|\Phi|}\{Q(\bar{F}_i), Q(\bar{G}_i), Q(\bar{H}_i)\} < Q$. Other constraints in MLCC_SL are of order $O(|\Phi| \cdot |B| \cdot T)$. The overall complexity of MLCC_L and MLCC_SL is $O(|\Phi| \cdot |B| \cdot T \cdot Q)$ and $O(|\Phi| \cdot |B| \cdot T + |\Phi| \cdot Q')$, respectively.

5.5 Performance Evaluation

In this section, we first introduce the simulation setting and then present the simulation results under various system parameters.

5.5.1 Simulation Setting

In our simulation setting, all the nodes in a network instance are distributed randomly within a 1000×1000 square region. The initial energy level of source nodes is specified as a Gaussian distribution with mean 1000 and variation 100. We set the variance of the background noise at each destination node and relay to 10^{-10}W.

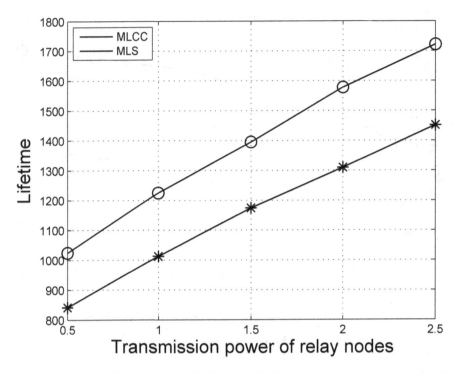

Fig. 5.7 Max-min lifetime versus transmission power of relay nodes

The channel bandwidth is specified as $W = 22$ MHz and the channel gain $|h_{xy}|^2$ between two nodes with a distance $||x - y||$ is calculated as $|h_{xy}|^2 = ||x - y||^{-4}$. The transmission rate requirement of each D2D pair is a Gaussian distribution with mean 1 and variation 0.1. For comparison, in addition to the DT scheme, we extend the MLS (Maximal Lifetime Scheduling) algorithm proposed in [26] to the channel-constrained network considered in this chapter. Its basic idea is to iteratively improve the minimal lifetime based on an initial assignment until it cannot be improved anymore. All results are averaged over 50 random network instances.

5.5.2 Simulation Results

We first investigate the effect of number of D2D pairs on the minimum lifetime under 10 relay nodes and 5 channels. The transmission power of relay nodes is set to one unit, i.e., $P_r = 1$, and the size of superframe is $T = 6$. As shown in Fig. 5.4, the minimum lifetime of multiple D2D pairs decreases as the pair number grows for both schemes and the results of MLCC always outperform MLS with larger lifetime. For example, the minimum lifetime is 1534 and 1209 under MLCC and MLS, respectively, in network instances with 11 D2D pairs. When the number of

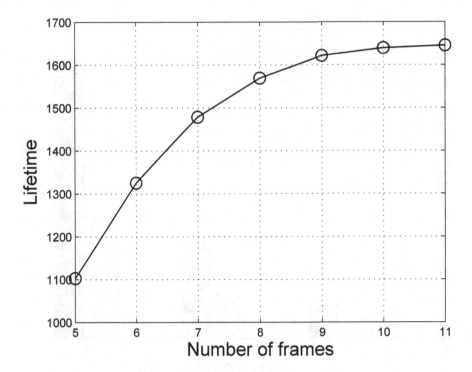

Fig. 5.8 Max-min lifetime versus number of frames in each transmission cycle

pairs grows to 19, the lifetime of MLCC and MLS decreases by 35.8 and 49.5 %, respectively. That is because more D2D pairs will share a channel in larger networks. In order to achieve the required transmission rate, they have to use higher transmission power to achieve the required communication rate. We also observe that DT performs the worst.

We then evaluate the lifetime performance under different number of relay nodes. The number of D2D pairs and channels are fixed to 15 and 5, respectively. As shown in Fig. 5.5, the relay node number does not affect the performance of DT, and the minimum lifetime of multiple D2D pairs under MLCC can be improved in the networks with more relay nodes. The lifetime performance of MLCC always outperforms MLS and increases from 1185 to 1758 as the number of relay nodes grows from 11 to 19. We attribute this phenomenon to the fact that each D2D pair has more chances to select a better relay node when larger number of relays are available in the network.

The influence of channel number to lifetime performance is investigated by changing its value from 3 to 8 in networks with 15 D2D pairs and 15 relay nodes. As shown in Fig. 5.6, the performance of all schemes increases as channel number grows. That is because the channel contention will be mitigated when more channels are available in the network. When the number of channels increases from 3 to 8, the lifetime improvement of MLCC and MLS is 1.38 and 1.44 times, respectively.

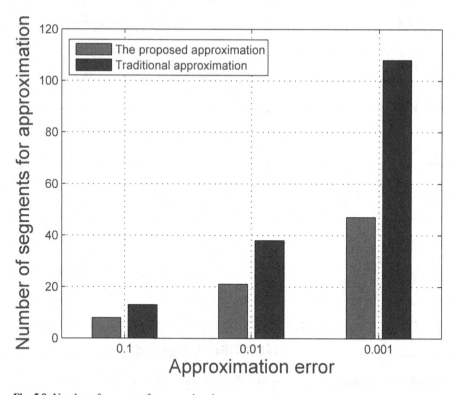

Fig. 5.9 Number of segments for approximation

The lifetime performance of both MLCC and MLS is also affected by the transmission power P_r of dedicated relay nodes. The DT scheme is not considered here because no relays are involved. To study this influence, we set P_r to different values and show the results in Fig. 5.7. The lifetime performance increases from 1024 to 1721 as the value of P_r grows from 0.5 to 2.5 under MLCC. That is because source nodes can use lower transmission power to achieve the rate requirements under larger value of P_r. In other words, the larger the value of P_r is set to, the more the portions of energy consumption are shifted from source nodes to relay nodes.

We study the influence of the number of time frames T in each transmission cycle by changing its value from 5 to 9 in networks with 15 D2D pairs, 15 relay nodes and 5 channels. As shown in Fig. 5.8, lifetime performance increases as the value of T grows since the benefits of dynamic channel allocation and relay selection can be better exploited under larger value of T. We obtain 19 % improvement of lifetime by increasing T from 5 to 6. Such gain will be quickly saturated, e.g., the lifetime only grows 3 % when T is increased from 8 to 9.

We compare the performance of Algorithm 4 and the traditional method shown in Fig. (5.2), in terms of number of linear constraints generated to approximate an exponential function, under various error bounds. As shown in Fig. 5.9, there are 8 and 13

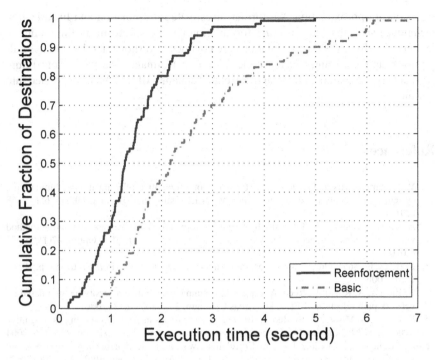

Fig. 5.10 Distribution of execution time

line segments in the proposed and traditional methods, respectively, when the approximation error is set to $\xi = 0.1$. As we set ξ to 0.001, the number of segments produced by our proposed method is 47, which is only 43.5 % of that using traditional method.

Finally, we investigate the execution time of the basic algorithm and reinforced one in network instances with 10 D2D pairs, 5 relay nodes and 3 channels. The values of T and P_r are set to 6 and 1, respectively. As shown in Fig. 5.10, the reenforcement can significantly accelerate the speed of solving the MLCC problem. For example, the number of execution time is less than 1 s in only 5 % network instances in the basic solution while the reenforcement can increase this value to 30 %. We also observe that the maximum execution time of basic solution and reenforcement is about 5 and 7 s, respectively.

5.6 Conclusion and Future Work

In this chapter, we study the energy efficiency of multiple D2D pairs with communication rate constraints using CC in multi-channel wireless networks. We formulate the MLCC problem in a MINLP form and prove it NP-hard. To solve this problem based on a branch-and-bound approach, we transform it into an MILP formulation using linear approximation. We further investigate reformulation and linearization techniques such that the resulting formulation can be solved in an time-efficient manner. Finally, extensive simulations are conducted to show that the proposed algorithm can significantly increase the lifetime performance. As part of our future work, energy-constrained relay and correlated channel [6] shall be incorporated into our model.

References

1. V. Asghari, S. Aissa, End-to-end performance of cooperative relaying in spectrum-sharing systems with quality of service requirements. IEEE Trans. Veh. Technol. **60**(6), 2656–2668 (2011)
2. A. Beck, A. Ben-Tal, L. Tetruashvili, A sequential parametric convex approximation method with applications to nonconvex truss topology design problems. J. Glob. Optim. **47**(1), 29–51 (2010)
3. E. Beres, R. Adve, On selection cooperation in distributed networks, in *IEEE CISS*, pp. 1056–1061, 2006
4. A. Bletsas, A. Khisti, D. Reed, A. Lippman, A simple cooperative diversity method based on network path selection. IEEE J. Sel. Areas Commun. **24**(3), 659–672 (2006)
5. D. Cabric, S. Mishra, R. Brodersen, Implementation issues in spectrum sensing for cognitive radios, in *IEEE Asilomar Conference on Signals, Systems and Computers*, pp. 772–776, 2004
6. A. Cacciapuoti, I. Akyildiz, L. Paura, Correlation-aware user selection for cooperative spectrum sensing in cognitive radio ad hoc networks. IEEE J. Sel. Area. Commun. **30**(2), 297–306 (2012)
7. X. Deng, A. Haimovich, Power allocation for cooperative relaying in wireless networks. IEEE Commun. Lett. **9**(11), 994–996 (2005)
8. W. Fang, F. Liu, F. Yang, L. Shu, S. Nishio, Energy-efficient cooperative communication for data transmission in wireless sensor networks. IEEE Trans. Consumer Electron. **56**(4), 2185–2192 (2010)
9. M. Felegyhazi, M. Cagalj, S. Bidokhti, J.-P. Hubaux, Non-cooperative multi-radio channel allocation in wireless networks, in *Proceeding IEEE INFOCOM*, pp. 1442–1450, May 2007
10. L. Gao, X. Wang, A game approach for multi-channel allocation in multi-hop wireless networks, in *Proceeding ACM MobiHoc*, pp. 303–312, 2008
11. C. Gao, Y. Shi, Y. Hou, H. Sherali, H. Zhou, Multicast communications in multi-hop cognitive radio networks. IEEE J. Select. Area. Commun. **29**, 784–793 (2011)
12. X. Gong, W. Yuan, W. Liu, W. Cheng, S. Wang, A cooperative relay scheme for secondary communication in cognitive radio networks, in *IEEE GLOBECOM*, pp. 1–6, 2008
13. M. Hajiaghayi, M. Dong, B. Liang, Optimal channel assignment and power allocation for dual-hop multi-channel multi-user relaying, in *IEEE INFOCOM*, pp. 76–80, April 2011
14. C. He, Z. Feng, Q. Zhang, Z. Zhang, H. Xiao, A joint relay selection, spectrum allocation and rate control scheme in relay-assisted cognitive radio system, in *IEEE VTC*, pp. 1 –5, 2010
15. X. Huang, H. Shan, X. Shen, On energy efficiency of cooperative communications in wireless body area network, in *IEEE WCNC*, pp. 1097–1101, March 2011
16. W. Ji, B. Zheng, Energy efficiency based cooperative communication in wireless sensor networks, in *IEEE ICCT*, pp. 938–941, Nov 2010
17. C. Jiang, Y. Shi, Y. Hou, W. Lou, Cherish every joule: maximizing throughput with an eye on network-wide energy consumption, in *Proceedings of IEEE INFOCOM*, pp. 1934–1941, March 2012
18. K. Khalil, M. Karaca, O. Ercetin, E. Ekici, Optimal scheduling in cooperate-to-join cognitive radio networks, in *IEEE INFOCOM*, pp. 3002–3010, April 2011

19. J. Laneman, D. Tse, G. Wornell, Cooperative diversity in wireless networks: efficient protocols and outage behavior. IEEE Trans. Inf. Theory **50**(12), 3062–3080 (2004)
20. Y. Li, P. Wang, D. Niyato, W. Zhuang, A dynamic relay selection scheme for mobile users in wireless relay networks, in *IEEE INFOCOM*, pp. 256–260, April 2011
21. P. Li, S. Guo, W. Zhuang, B. Ye, Capacity maximization in cooperative crns: joint relay assignment and channel allocation, in *IEEE ICC*, pp. 6619–6623, March 2012
22. P. Li, S. Guo, V. Leung, Improving throughput by fine-grained channel allocation in cooperative wireless networks, in *IEEE GLOBECOM*, pp. 5740–5744, 2012
23. P. Li, S. Guo, Z. Cheng, A.V. Vasilakos, Joint relay assignment and channel allocation for energy-efficient cooperative communications, in *IEEE WCNC*, pp. 5740–5744, 2012
24. P. Li, S. Guo, Y. Xiang, H. Jin, Unicast and broadcast throughput maximization in amplify-and-forward relay networks. IEEE Trans. Veh. Technol. **61**(6), 2768–2776 (2012)
25. P. Li, S. Guo, W. Zhuang, B. Ye, On efficient resource allocation for cognitive and cooperative communications, in *Accepted by IEEE Journal on Selected Areas in Communications*, 2013
26. G. Liu, L. Huang, H. Xu, W. Liu, Y. Zhang, Maximal lifetime scheduling for cooperative communications in wireless networks, in *Proceeding International Conference on Computer Communications and Networks (ICCCN)*, pp. 1–6, 2010
27. S. Mishra, A. Sahai, R. Brodersen, Cooperative sensing among cognitive radios, in *IEEE ICC*, vol. 4, pp. 1658–1663, June 2006
28. C.H. Papadimitriou, K. Steiglitz, *Combinatorial Optimization: Algorithms and Complexity* (Prentice-Hall, Inc., Upper Saddle River, 1982)
29. S. Sharma, Y. Shi, Y.T. Hou, S. Kompella, An optimal algorithm for relay node assignment in cooperative ad hoc networks, in *IEEE/ACM Transactions on Networking*, pp. 879–892, 2010
30. S. Sharma, Y. Shi, Y.T. Hou, H.D. Sherali, S. Kompella, Cooperative communications in multi-hop wireless networks: joint flow routing and relay node assignment, in *IEEE INFOCOM*, pp. 1–9, 2010
31. S. Sharma, Y. Shi, J. Liu, Y. Hou, S. Kompella, Is network coding always good for cooperative communications? in *IEEE INFOCOM*, pp. 1–9, March 2010
32. S. Sharma, Y. Shi, Y. Hou, H. Sherali, S. Kompella, Optimizing network-coded cooperative communications via joint session grouping and relay node selection, in *IEEE INFOCOM*, pp. 1898–1906, April 2011
33. T.-C. Shih, C.-C. Kao, S.-R. Yang, A cooperative mac protocol in multi-channel wireless ad hoc networks, in *Proc. Wireless Communications and Mobile Computing Conference (IWCMC), 2011 7th International*, pp. 1831–1836, 2011.
34. L. Simic, S. Berber, K. Sowerby, Energy-efficiency of cooperative diversity techniques in wireless sensor networks, in *IEEE PIMRC*, pp. 1–5, Sept 2007
35. E. C. van der Meulen, Three-terminal communication channels, Adv. Appl. Probab. **3**, 120–154 (1971)
36. F. Wu, S. Zhong, C. Qiao, Globally optimal channel assignment for non-cooperative wireless networks, in *Proc. IEEE INFOCOM*, pp. 1543 –1551, April 2008
37. F. Wu, N. Singh, N. Vaidya, G. Chen, On adaptive-width channel allocation in non-cooperative, multi-radio wireless networks, in *Proceeding IEEE INFOCOM*, pp. 2804–2812, April 2011
38. D. Yang, X. Fang, G. Xue, Hera: an optimal relay assignment scheme for cooperative networks. IEEE J. Select. Area. Commun. **30**(2), 245–253 (2012)
39. J. Zhang, Q. Zhang, Stackelberg game for utility-based cooperative cognitive radio networks, in *ACM MobiHoc*, pp. 23–32, 2009
40. Y. Zhao, R. Adve, T. Lim, Improving amplify-and-forward relay networks: optimal power allocation versus selection. IEEE Trans. Wirel. Commun. **6**(8), 3114–3123 (2007)

Chapter 6
Cooperative Device-to-Device Communication for Broadcast

Abstract Cooperative D2D communication offers an efficient and low-cost way to achieve spatial diversity by forming a virtual antenna array among single-antenna nodes that cooperatively share their antennas. It has been well recognized that the selection of relay nodes plays a critical role in the performance of cooperative D2D communication. Most existing relay selection strategies focus on optimizing the outage probability or energy consumption. To fill in the vacancy of research on throughput improvement via cooperative communication, we study the relay selection problem with the objective of optimizing the throughput in this chapter. For unicast, it is a P problem and an optimal relay selection algorithm is provided with a correctness proof. For broadcast, we show the challenge of relay selection by proving it NP-hard. A greedy heuristic algorithm is proposed to effectively choose a set of relay nodes that maximize the broadcast throughput. Simulation results show that the proposed algorithms can achieve high throughput under various network settings.

6.1 Introduction

Spatial diversity has shown supreme effectiveness in combating multipath fading in wireless environments by employing multiple transceiver antennas, such as multiple-input multiple-output (MIMO) techniques. However, it is not always practical to equip a wireless node with multiple antennas due to the size and cost constraints of wireless devices, e.g., cellphones and sensors. Instead of requiring multiple transceiver antennas on the same node, cooperative communication (CC) offers an efficient and low-cost way to achieve spatial diversity by forming a virtual antenna array among single-antenna nodes that cooperatively share their antennas.

The essence of cooperative D2D communication is to exploit the wireless broadcast advantage and the relaying capability of neighboring nodes so as to achieve higher throughput and lower transmission error rate. Typically, there are two modes: AF (Amplify-and-Forward) and DF (Decode-and-Forward) in cooperative communications. Under AF, the cooperative relay node amplifies the signal received from the source and then forwards it to the destination node. Under DF, the cooperative relay node decodes the received signal and re-encodes it before forwarding it to the destination node. Due to the lower implementation complexity of AF, we focus our node cooperation strategy on the AF mode throughout this chapter.

© The Author(s) 2014   61
P. Li, S. Guo, *Cooperative Device-to-Device Communication*
in Cognitive Radio Cellular Networks, SpringerBriefs in Computer Science,
DOI 10.1007/978-3-319-12595-4_6

It has been shown that cooperative D2D communication can improve network reliability and energy consumption [12, 14, 17, 20]. In particular, the selection of relay nodes plays a critical role in the performance of cooperative D2D communication. For example, for a single D2D link, the full diversity order can be achieved by choosing the "best" relay node [26]. Most existing relay selection strategies focus on optimizing the outage probability or energy consumption [5, 8, 10, 19]. However, many applications in wireless networks desire high throughput provided by cooperative communication, which motivates us to explore the potential of throughput improvement via relay selection.

In this chapter, we study the relay selection problem with the objective of optimizing the throughput. For unicast problem, traditional best relay selection algorithm [26] explores the full diversity order but fails to guarantee the maximum throughput. We then develop an optimal algorithm, a correctness proof, that achieves the maximum unicast throughput by selecting a number of relay nodes. For the more challenging broadcast problem that has not been investigated so far, we formulate it as a max-min problem. The simulation results show that the proposed CC-based algorithms can significantly improve the throughput over the direct transmission protocol under various network settings.

The rest of this chapter is organized as follows. Section 6.2 presents system model. The relay selection problems for unicast and broadcast are studied in Sections 6.3 and 6.4, respectively. Section 6.5 gives the simulation results of performance evaluation. Finally, Section 6.6 concludes the chapter with a summary and outlook on future work.

6.2 System Model

We consider a stationary AF relay network in which a source s disseminates data to a set of destinations $N_d = \{d_1, d_2, ..., d_{n_d}\}$ with the support of a dedicated set of relay nodes $N_r = \{r_1, r_2, ..., r_{n_r}\}$. The source node could be a base station or a mobile user. Note that in unicast, N_d contains only one destination, i.e., $n_d = 1$. Suppose a relay node set $M \subseteq N_r$ to be selected to assist the transmission, in which a frame is divided into $|M| + 1$ time slots. Note that when $M = \emptyset$, it degenerates to the direction transmission case. In the first time slot, source node s broadcasts a packet, which is overheard by all the relays and the destinations in its transmission range. Then, the selected relays in M amplify and forward the packet one by one in the subsequent $|M| = m$ time slots. Note that all the nodes in the network work in a half-duplex mode in which they cannot transmit and receive simultaneously. Therefore, the channel capacity between s and d with the relay set M can be expressed as:

$$C(s, M, d) = \frac{W}{|M| + 1} \log_2 \left(1 + \gamma_{sd} + \sum_{r_i \in M} F(\gamma_{sr_i}, \gamma_{r_i d}) \right) \qquad (6.1)$$

In (6.1), we observe that the channel capacity is determined by two factors. One is the SNR contribution of each selected relay node and the other one is the number of selected relay nodes m. On the one hand, multiple relay nodes involving in the same cooperative communication session is desired since they can contribute more SNR with an increased channel capacity. On the other hand, too many relay nodes would degrade the overall performance due to the time division multiplexing over the same bandwidth. This observation motivates us to make a good tradeoff between these two factors in the relay selection.

6.3 Relay Selection for Unicast

We consider an amplify-and-forward relay network with a source s that transmits information to a destination d. The transmission rate with a given relay node set $M \subseteq N_r$ is therefore

$$R_u(M) = C(s, M, d). \tag{6.2}$$

When the relay set M is empty, the unicast throughput becomes the direct transmission rate. The objective of the maximum throughput unicast problem in AF relay networks (MTU-AF) is to find the optimal relay node set M^* that maximizes the transmission rate of source s, i.e.,

$$M^* = \arg \max_{M \subseteq N_r} \{R_u(M)\}. \tag{6.3}$$

We propose a polynomial-time optimal relay selection algorithm for unicast (ORSU) to maximize the throughput. The pseudo code of ORSU is given in Algorithm 1. At the beginning, the selected relay node set M is initialized to be empty and the corresponding throughput $curr\_rate$ is given the value of direct transmission rate $C(s, \emptyset, d)$. Then, we use **sort_F** procedure in line 3 to sort the relay nodes $r_i \in N_r$ to $\mathcal{N}$ according to their results of $F\left(\gamma_{sr_i}, \gamma_{r_id}\right)$ in a descending

---

Algorithm 5 The optimal relay selection algorithm for unicast (ORSU)

1: $M = \emptyset$
2: $curr\_rate = C(s, \emptyset, d)$
3: $\mathcal{N} = \textbf{sort\_F}(N_r)$
4: **while** $i = 1$ to n_r **do**
5:     $M' = \mathcal{N}(1 : i)$
6:     $rate = R_u(M')$
7:     **if** $rate > curr\_rate$ **then**
8:         $M = M'$
9:         $curr\_rate = rate$
10:    **end if**
11: **end while**
12: **return** M

---

order. In each iteration of **while** loop, we estimate the transmission rate using the first i elements in $\mathcal{N}$ as the relay nodes. If it is greater than $curr\_rate$, we update M as well as $curr\_rate$ to the current best results. Finally, we return the relay node set M with maximum transmission rate. It is a straightforward exercise to show that the computational complexity of the ORSU algorithm is dominated by the procedure **sort_F**, which can be finished within $O(n_r \log n_r)$ by using quick sort.

Theorem 6.1 *The solution found by the ORSU algorithm is optimal.*

Proof We prove this theorem by contradiction. Suppose there exist an optimal relay node set M', which is different from M found by ORSU with $R_u(M') > R_u(M)$. Since our ORSU algorithm checks all transmission rates with various relay sets that include the first contiguous i nodes in the sorted set $\mathcal{N}$, there must exist two relays r_i and r_j, where i and j are the corresponding indices in $\mathcal{N}$, satisfying $i < j$, $r_i \notin M'$ and $r_j \in M'$. By replacing r_j with r_i in M', a new relay set is formed with a non-decreased transmission rate. We repeat the same operation for all such possible replacements and eventually obtain a relay set M'', in which all relay nodes are contiguous in the front of $\mathcal{N}$. According to our ORSU algorithm, we conclude $R_u(M) \geq R_u(M'') \geq R_u(M')$, which contradicts with the assumption.

6.4 Relay Selection for Broadcast

We study the relay selection problem in a broadcast session with a source node s and a set of associated destinations N_d. Similar to unicast, a set of relay nodes $M \subseteq N_r$ are selected to assist the broadcast session. Note that a relay node can serve several destinations simultaneously due to the broadcast characteristic of wireless channel. In order to guarantee a packet to be received by all destinations, the broadcast throughput R_b should not exceed the capacity of any channel between source s and a destination in N_d, i.e.,

$$R_b(M) = \min_{d_i \in N_d} \{C(s, M, d_i)\}. \tag{6.4}$$

The objective of the maximum throughput broadcast problem in AF relay networks (MTB-AF) is to find the optimal relay node set that maximizes the broadcast throughput, i.e.,

$$M^* = \arg \max_{M \subseteq N_r} \{R_b(M)\} = \arg \max_{M \subseteq N_r} \min_{d_i \in N_d} \{C(s, M, d_i)\}. \tag{6.5}$$

Based on the above formulation, we have two observations. First, the optimal relay node set M^* could be empty, which implies the direct transmission. Second, it is not required to have all destinations covered by relay nodes in M^* since the coverage requirement may lead to a large denominator in formula (6.1), which is not desirable for throughput performance.

Fig. 6.1 A simple example of cooperative broadcast

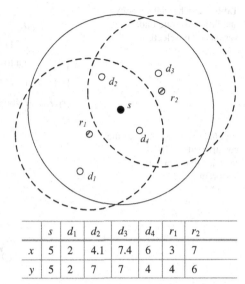

Table 6.1 The locations of nodes in Fig. 6.1

	s	d_1	d_2	d_3	d_4	r_1	r_2
x	5	2	4.1	7.4	6	3	7
y	5	2	7	7	4	4	6

6.4.1 Complexity Analysis

A straightforward method to solve the MTB-AF problem is the approach to extend the ORSU algorithm for broadcast, denoted as BE_ORS, which combines the relay sets obtained by ORSU for each destination. Unfortunately, the broadcast throughput under this simple extension is not satisfactory and sometimes even lower than the direct transmission rate since too many relays would be selected. In order to maximize the broadcast throughput, we need to jointly consider the number of relay nodes and their SNR contributions.

The challenges of the MTB-AF problem can be illustrated by the simple example in Fig. 6.1, where four destination nodes $\{d_1, d_2, d_3, d_4\}$ and two relay nodes $\{r_1, r_2\}$ are all in the transmission range of a source node s. Their locations are given in Table 6.1. Under the same system parameter settings given in Section VI, we calculate the channel capacity and broadcast rate for all possible relay selection strategies as summarized in Table 6.2. The results of direct transmission are shown in first row with empty M. We observe that the optimal solution is $M = \{r_1\}$, in which d_3 is not covered by any relay node. We also notice that the selection of both relay nodes $\{r_1, r_2\}$ leads to a lower broadcast rate because more relays take up the shared bandwidth, even though more SNR contributions are provided to all destinations.

The above exhaustive search approach is obviously not scalable in large-scale network topologies because of the exponential combinations of relay node selection. Finally, we show that the MTB-AF problem is intractable by proving it NP-hard in the following theorem.

Table 6.2 Channel capacity and broadcast rate under various relay selections

M	d_1	d_2	d_3	d_4	R_b
$\emptyset$	2.5447	18.8807	7.2780	40.5528	2.5447
$\{r_1\}$	4.4407	9.8149	3.6390	20.3833	3.6390
$\{r_2\}$	1.2724	9.8734	9.6843	20.5564	1.2724
$\{r_1, r_2\}$	2.9605	6.8177	6.4562	13.7728	2.9605

Fig. 6.2 An illustration of the constructed MTB-AF instance in the proof of Theorem 2

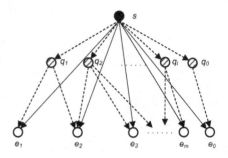

Theorem 6.2 *The MTB-AF problem is NP-hard.*

Proof In order to prove the NP-hardness of an optimization problem, we need to show its decision form to be NP-complete. In the following, we first formalize the MTB-AF problem in its decision form:

The MTB-AF problem

INSTANCE: Given a source node s, a set of destination nodes $N_d = \{d_1, d_2, ..., d_{n_d}\}$, a set of relay nodes $N_r = \{r_1, r_2, ..., r_{n_r}\}$ and a positive constant $X \in R^+$.

QUESTION: Is there a set of relay nodes $M \subseteq N_r$ such that the resulting broadcast throughput $R_b(M) \geq X$?

The MTB-AF problem is clearly in NP class as the objective function associated with a given selected relay node set can be evaluated in a polynomial time. The remaining proof is done by reducing the well-known NP-complete set cover (SC) problem to MTB-AF.

The SC problem

INSTANCE: Given a set of elements $E = \{e_1, e_2, ..., e_m\}$, a set of subsets $Q = \{q_1, q_2, ..., q_l\}$ of E (i.e., $q_i \subseteq E$, $1 \leq i \leq l$), and a constant $Y \in R^+$.

QUESTION: Is there a collection Ψ of sets from Q that covers all elements in E, i.e., $\bigcup_{q_i \in \Psi} q_i = E$, such that $|\Psi| \leq Y$?

We now describe the transformation of an SC instance into an instance of MTB-AF problem as shown in Fig. 6.2, in which sets E and Q correspond to destination set and relay set, respectively. In addition, a virtual relay node q_0 and a virtual destination node e_0 in its transmission range are also included into the network. Finally, a node s is attached to the network as the broadcast source. We set the channel capacity under direct transmission $C(s, \emptyset, e_i) < X$ for $0 \leq i \leq m$ shown as solid arrows in Fig. 6.2. We define function F as $F(\gamma_{sq_j}, \gamma_{q_j e_i}) = f_0 > 0$ if $e_i \in q_j$ is satisfied in the original SC problem, and $F(\gamma_{sq_j}, \gamma_{q_j e_i}) = 0$ otherwise. We further set $F(\gamma_{sq_0}, \gamma_{q_0 e_0}) = f_0$

and $W \log_2 \left(1 + \gamma_{se_i} + f_0\right) = c_0$ for $0 \leq i \leq m$. By such settings, only relay links (s, q_j) and (q_j, e_i) with a positive F value are represented by dotted arrows in Fig. 6.2. The above construction process can be done in a straightforward way and in polynomial time by choosing proper values for γ_{se_i}, γ_{sq_j} and $\gamma_{q_j e_i}$.

To prove the MTB-AF problem to be NP-complete, we only need to show that SC problem has a solution if and only if the resulting instance of MTB-AF has a set of relay nodes satisfying the broadcast throughput requirement. This can be achieved by setting

$$X \equiv \frac{c_0}{Y + 2}. \tag{6.6}$$

For the only-if case, we suppose that there exists a collection of subsets Ψ covering all the elements in E and $|\Psi| \leq Y$. Then for the MTB-AF problem, we set $M = \Psi \cup \{q_0\}$ and the resulting broadcast throughput can be calculated as: $R_b(M) = \min_{d_i \in E \cup \{e_0\}} \{C(s, M, d_i)\} = C\left(s, \{q_0\}, e_0\right) = \frac{c_0}{|M|+1} = \frac{c_0}{|\Psi|+2} \geq \frac{c_0}{Y+2} = X$.

For the if case, we suppose that an instance of MTB-AF problem has a relay node set M with $R_b(M) \geq X$. Note that q_0 must be included in M under this condition and the corresponding solution of SC problem is $\Psi = M - \{q_0\}$. We claim that Ψ is a cover of E since if any destination node, e.g., e_i, has not been included in Ψ, the resulting broadcast throughput $R_b(M) = \frac{C(s, \emptyset, e_i)}{|M|+1} < \frac{X}{|M|+1} \leq X$, which contradicts $R_b(M) \geq X$. Finally, from the derivation $\frac{c_0}{Y+2} = X \leq R_b(M) = C\left(s, \{q_0\}, e_0\right) = \frac{c_0}{|M|+1} = \frac{c_0}{|\Psi|+2}$, we obtain $|\Psi| \leq Y$.

To study the approximate solutions of the MTB-AF problem, we have the following discovery.

Theorem 6.3 *There is no polynomial-time algorithm with constant approximation ratio for the MTB-AF problem unless $P = NP$.*

Proof We achieve this conclusion by contradiction. Let $(\cdot)^*$ denote an optimal solution. Suppose that for some constant number ρ, there is a polynomial-time algorithm $\mathcal{X}$ with approximation ratio $\rho > 1$ for the MTB-AF problem, i.e.,

$$\rho \cdot X \geq X^*. \tag{6.7}$$

The polynomial-time transformation process in the proof of Theorem 2 will lead to the result that algorithm $\mathcal{X}$ is a constant-factor approximation algorithm for the SC problem as well. This is explained as follows. Because of (6.6), (6.7) and the fact of $Y^* \geq 1$, we have $Y = \frac{c_0}{X} - 2 \leq \frac{\rho \cdot c_0}{X^*} - 2 = \rho \cdot Y^* + 2(\rho - 1) \leq \rho \cdot Y^* + 2(\rho - 1) \cdot Y^* = (3\rho - 2) \cdot Y^*$. In other words, $\mathcal{X}$ is a $(3\rho - 2)$-factor approximation algorithm for the SC problem, which is contradicted to the fact in [1] that there is no polynomial-time algorithm with constant approximation ratio for the SC problem unless $P = NP$.

6.4.2 A Greedy Relay Selection Algorithm for Broadcast

Due to the NP-hardness and inapproximability of the MTB-AF problem, we focus on the design of an effective heuristic algorithm formulated in Algorithm 2 to maximize the broadcast throughput. The basic idea of our greedy relay selection algorithm for broadcast (GRSB) is to iteratively include one more relay node that leads to the maximum broadcast throughput until it cannot be further increased.

Algorithm 6 The greedy relay selection algorithm for broadcast (GRSB)

1: $M = \emptyset$
2: $curr\_rate = 0$
3: **while** $M \neq N_r$ **do**
4:     $r = \arg\max_{r' \in N_r - M} R_b(M \cup \{r'\})$
5:     $rate = R_b(M \cup \{r\})$
6:     **if** $rate > curr\_rate$ **then**
7:        $M = M \cup \{r\}$
8:        $curr\_rate = rate$
9:     **else**
10:        break
11:     **end if**
12: **end while**
13: **return** M

In Algorithm 2, the set of selected relay nodes M is initialized to be empty and the corresponding broadcast throughput $curr\_rate$ is set to zero at the beginning. Then, the algorithm proceeds by adding relay nodes to M in **while** loop from line 3 to line 12. In each iteration of **while** loop, the best relay node r is chosen such that the resulting broadcast throughput $rate$ is maximized. If $rate$ is greater than $curr\_rate$, r is added to M. Otherwise, the current M is returned. The time complexity of the GRSB algorithm is $O(n_r^2 n_d)$ that can be explained as follows. The **while** loop performs at most n_r iterations. In each iteration of **while** loop, the major calculation in line 4 can be updated in $O(n_r n_d)$.

6.5 Performance Evaluation

In our simulation, the bandwidth for wireless channels is set as $W=10$ MHz, the transmission power to noise ratio ϕ is set the same for all transmitting nodes, and the path-loss component between nodes u and v is given by $||u - v||^{-4}$, in which $||u - v||$ is the distance between these two nodes. All the nodes in the network are placed within a 10×10 square region. For unicast, the source and destination are fixed at $(2, 5)$ and $(8, 5)$, respectively, and relay nodes are randomly distributed in the region. For broadcast, all the nodes including source, destinations and relays are randomly distributed in the square region. Under each setting with a specific n_r and a given ϕ, all simulation results in the following are obtained by averaging over 1000 network instances in a PC with Intel Core2 Duo 2.53 GHz CPU.

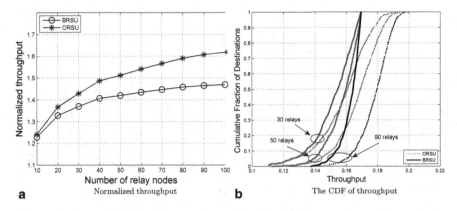

Fig. 6.3 The performance under different different number of relay nodes for unicast

6.5.1 Unicast

We compare the performance of the ORSU algorithm with the best relay selection algorithm for unicast [26] , referred as BRSU, under various numbers n_r of relay nodes from 10 to 100. The performance is evaluated in terms of normalized throughput which is defined as the ratio of transmission rate of ORSU/BRSU to the rate of direct transmission. Under a fixed $\phi = 10dB$, the experimental results given in Fig. 6.3a show that both algorithms have much higher throughput than direct transmission and the normalized throughput is an increasing function of the number of relay nodes n_r. Furthermore, the ORSU algorithm outperforms the BRSU algorithm, especially in the networks with more relay nodes. For example, while the normalized throughputs of ORSU and BRSU are similar, about 1.22, in the scenarios with 10 relay nodes, these numbers increase to 1.61 and 1.46, respectively, when 100 relay nodes are available. We also plot the CDF (Cumulative Distributed Function) of throughput measurements under 30, 60 and 90 relay nodes in Fig. 6.3b. Since the throughput of direct transmission does not change under a given network setting, we omit it in the unicast evaluation in Figs. 6.3b, 6.4b and 6.5b. The minimum values of all the curves are not less than the throughput of direct transmission whose value is 0.11. Moreover, the throughput is increased as the number of relay nodes grows for both algorithms. In particular, the maximum throughput of BRSU under various number of relay nodes merge to the value of 0.17, which is the up-bound of the BRSU algorithm in our simulation setting. In ORSU algorithm, nearly 30 %, 50 % and 90 % of throughput measurements under 30, 50 and 90 relay nodes, respectively, are greater than this up-bound.

Then, we fix the number of relay nodes to 60 and study the influence of transmission power to noise ratio to the throughput performance as shown in Fig. 6.4a. The advantages of ORSU and BRSU are more significant in the networks with larger value of ϕ. As ϕ grows from $10dB$ to $20dB$, the normalized throughput of ORSU is increased from 1.43 to 2.53, which is always better than BRSU. The difference

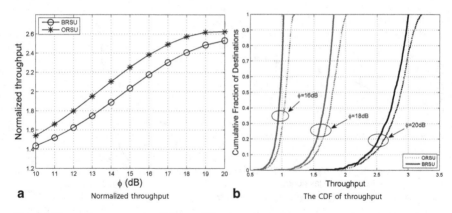

Fig. 6.4 The performance under different different transmission power-to-noise ratio for unicast

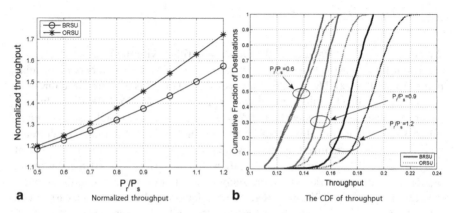

Fig. 6.5 The performance under different different transmission power of relay nodes for unicast

between ORSU and BRSU is small when ϕ is 10 because of the limited contribution of each relay node. When ϕ is increased to 14, the gap is maximized to 0.22. Further increase of ϕ cannot bring more advantages since the larger value of ϕ at source dominates the capacity of the unicast channel. The CDF of throughput measurements under different transmission power-to-noise ratio is shown in Fig. 6.4b. Obviously, the throughput of BRSU and ORSU shows an increasing function of SNR. This is because that large SNR indicates higher transmission power or lower noise level, both of which have a positive effect on the throughput. Moreover, ORSU always outperforms BRSU under various SNR values, which coincides with the observation in Fig. 6.4a.

Finally, we evaluate the throughput performance when relay nodes use different transmission power P_r from P_s by changing their ratio from 0.5 to 1.2. The number of relay nodes is set to 60 and transmission power to noise ratio is 10 dB. As shown in Fig. 6.5a, the normalized throughput is increased as the ratio grows for both algorithms. For example, the throughput of ORSU is 1.20 times of direct transmission

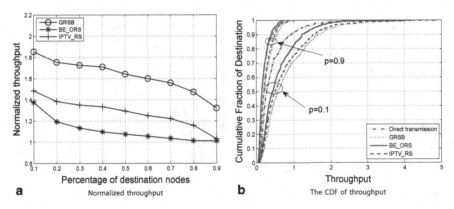

Fig. 6.6 The performance under different percentage of destination nodes for broadcast

when the power ratio is 0.5. Under the same scenario, BRSU outperforms direct transmission of 1.18 times. When the power ratio increased to 1.2, which means the transmission power of relay nodes is greater than source, the normalized throughput of ORSU and BRSU is grows to 1.72 and 1.57, respectively. We also observe the gap between them is more obvious under the bigger power ratio since each relay node can bring more improvement under larger transmission power. As shown in Fig. 6.5b, we plot the CDF of throughput under different transmission power of relay nodes for further comparison. The transmission power of relay nodes has significant influence to the throughput performance for both algorithms. The larger it is, the higher throughput can be obtained. When the power ratio is 0.6, the maximum throughput of ORSU can be achieved is 0.167. As the ratio grows to 0.9 and 1.2, over 40 and 95 % throughput measurements is higher than this value, respectively. We have the similar observation of the advantages brought by high transmission power of relay nodes in BRSU algorithm.

6.5.2 Broadcast

For broadcast, we evaluate three relay selection algorithms in our simulation. In addition to the GRSB algorithm and the BE_ORS algorithm, we consider the relay selection algorithm designed for IPTV networks in [18], which is referred to as IPTV_RS in our chapter. In IPTV_RS, the relay node covering the most number of destinations is selected in each step until the throughput does not increase any more. The total number of destination nodes and relay nodes is 100 in our simulation and we use p $(0 < p < 1)$ to denote the percentage of destination nodes. We first investigate the effect of node composition on the broadcast throughput by increasing p from 0.1 to 0.9. A small value of p indicates the scenario with a small number of destinations but much more relays, whereas large value represents that most of nodes in the network are destinations with only a few relays. The transmission power-to-noise ratio ϕ is fixed to 20 dB. As shown in Fig. 6.6, the performance of the GRSB

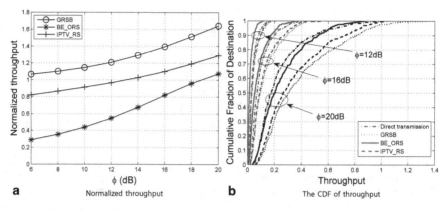

Fig. 6.7 The performance under different transmission power-to-noise ratio for broadcast

algorithm always outperforms other algorithms. For example, when $p = 0.1$, i.e., 10 destinations and 90 relays in the network, the broadcast throughput of GRSB achieves 1.85 times of the direct transmission, while only a little improvement is obtained by IPTV_RS and BE_ORS. On the other hand when both IPTV_RS and BE_ORS degrades to a similar performance of direction transmission under $p = 0.9$, the normalized performance of GRSB still achieves as high as 1.32. We also show the CDF of broadcast throughput under $p = 0.1$ and 0.9 in Fig. 6.6b. When a large number of randomly placed relay nodes are available, there is a higher probability to choose the ones that can bring more contribution to the throughput. In particular, almost all throughput measurements are lower than 1 under $p = 0.9$. However, when the value of p decreases to 0.1, the maximum throughput of GRSB can achieve 6.07. We also notice that the execution time of the GRSB algorithm is always less than 0.04s under various values of p, showing it a practical heuristic algorithm.

We then investigate the affect of transmission power-to-noise ratio to the throughput performance by changing its value from 6 to 20 dB under $p = 0.5$. As shown in Fig. 6.7a, GRSB always outperforms direct transmission and its normalized throughput increases from 1.06 to 1.64 as ϕ grows from 6 to 20 dB. However, the throughput of IPTV_RS and BE_ORS is lower than direct transmission when the value of ϕ is less than 14 to 20 dB, respectively. The CDF of throughput under different transmission power-to-noise ratio is shown in Fig. 6.7b. Similar with unicast, the broadcast throughput is an increasing function of ϕ for all the transmission schemes. The advantages of GRSB algorithm are significant especially under larger values of ϕ. For example, while the maximum throughput of GRSB is only 0.2 under $\phi = 14dB$, over 20 and 65 % of network instances with a broadcast throughput greater than this value can be achieved under $\phi = 16dB$ and $20dB$, respectively.

To evaluate the influence of relay nodes' transmission power to the throughput, we increase the transmission power ratio of source and relay nodes from 0.5 to 1.2 and show the results in Fig. 6.8a. When the ratio is 0.5, GRSB outperforms direct transmission by 1.15 times whereas the performance of IPTV_RS and BE_ORS is only 89.3 and 64.7 % of direct transmission, respectively. When the ratio grows to 1,

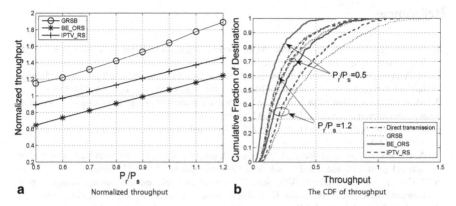

Fig. 6.8 The performance under different transmission power of relay nodes for broadcast

the normalized throughput of GRSB, IPTV_RS and BE_ORS increases to 1.64, 1.45 and 1.07, respectively. We show the CDF of throughput under different transmission power of relay nodes in Fig. 6.8b. There is only one curve to show the throughput of direct transmission since it is not affected by the transmission power of relay nodes. We have similar observation on Fig. 6.5b that the broadcast throughput can be increased under higher transmission power of relay nodes.

In summary, GRSB significantly outperforms other heuristic algorithms under various simulation settings. We attribute this phenomenon to the fact that our GRSB algorithm achieves a good balance on maximizing channel capacity while relying on only a small number of relay nodes. In contrast, both IPTV_RS and BE_ORS ignore the influence of relay node number on the broadcast throughput and select too many relay nodes to achieve a satisfactory performance.

6.6 Conclusion

In this chapter, we study the maximum throughput problem in amplify-and-forward relay networks for both unicast and broadcast. We focus on the relay selection scheme due to its significance on the performance. For unicast, it is a P problem and an optimal algorithm is proposed. For broadcast, the maximum throughput problem is proved to be NP-hard and a heuristic algorithm with good performance is proposed. As part of the future work, we shall take power control of relay nodes into consideration to develop energy efficient cooperative broadcast protocol.

References

1. N. Alon, D. Moshkovitz, S. Safra, Algorithmic construction of sets for k-restrictions. ACM Trans. Algorithm **2**(2), 153 – 177 (2006)
2. K. Azarian, H. El Gamal, P. Schniter, On the achievable diversity-multiplexing tradeoff in half-duplex cooperative channels. IEEE Trans. Inf. Theory **51**(12), 4152 –4172 (2005)
3. M. Baghaie, B. Krishnamachari, Delay constrained minimum energy broadcast in cooperative wireless networks, in *IEEE INFOCOM*, pp. 864 –872, 2011
4. E. Beres, R. Adve, On selection cooperation in distributed networks, in *IEEE CISS*, pp. 1056 –1061, 2006
5. A. Bletsas, A. Khisti, D. Reed, A. Lippman, A simple cooperative diversity method based on network path selection. IEEE J. Sel. Area Commun. **24**(3), 659 – 672 (2006)
6. J. Cai, X. Shen, J. Mark, A. Alfa, Semi-distributed user relaying algorithm for amplify-and-forward wireless relay networks. IEEE Trans. Wirel. Commun. **7**(4), 1348 –1357 (2008)
7. M. Čagalj, J. Hubaux, C. Enz, Minimum-energy broadcast in all-wireless networks: Np-completeness and distribution issues, in *ACM MobiCom*, pp. 172–182, 2002
8. X. Deng, A. Haimovich, Power allocation for cooperative relaying in wireless networks. IEEE Commun. Lett. **9**(11), 994 – 996 (2005)
9. D. Gunduz, E. Erkip, Opportunistic cooperation by dynamic resource allocation. IEEE Trans. Wirel. Commun. **6**(4), 1446 –1454 (2007)
10. A. Host-Madsen, J. Zhang, Capacity bounds and power allocation for wireless relay channels. IEEE Trans. Inf. Theory **51**(6), 2020 –2040 (2005)
11. A. Khandani, J. Abounadi, E. Modiano, L. Zheng, Cooperative routing in static wireless networks. IEEE Trans. Commun. **55**(11), 2185 –2192 (2007)
12. J. Laneman, D. Tse, G. Wornell, Cooperative diversity in wireless networks: efficient protocols and outage behavior. IEEE Trans. Inf. Theory **50**(12), 3062 – 3080 (2004)
13. P. Li, S. Guo, W. Zhuang, B. Ye, Capacity Maximization in cooperative CRNs: joint relay assignment and channel allocation, in *IEEE ICC* 2012
14. I. Maric, R. Yates, Cooperative multihop broadcast for wireless networks. IEEE J. Sel. Areas Commun. **22**(6), 1080 – 1088 (2004)
15. R. Nabar, H. Bolcskei, F. Kneubuhler, Fading relay channels: performance limits and space-time signal design. IEEE J. Sel. Areas Commun. **22**(6), 1099 – 1109 (2004)
16. T. C.-Y. Ng, W. Yu, Joint optimization of relay strategies and resource allocations in cooperative cellular networks. IEEE J. Sel. Areas Commun. **25**(2), 328 –339 (2007)
17. A. Ribeiro, X. Cai, G. Giannakis, Symbol error probabilities for general cooperative links. IEEE Trans. Wirel. Commun. **4**(3), 1264 – 1273 (2005)
18. B. Rong, A. Hafid, Cooperative multicast for mobile iptv over wireless mesh networks: the relay-selection study. IEEE Trans. Veh. Technol. **59**(5), 2207 –2218 (2010)
19. A. Scaglione, D. Goeckel, J. Laneman, Cooperative communications in mobile ad hoc networks. IEEE Signal Process Mag. **23**(5), 18 – 29 (2006)
20. S. Sharma, Y. Shi, Y. T. Hou, S. Kompella, An optimal algorithm for relay node assignment in cooperative ad hoc networks. IEEE/ACM Trans. Netw. 99, 1 (2010)
21. S. Sharma, Y. Shi, Y. T. Hou, H. D. Sherali, S. Kompella, Cooperative communications in multi-hop wireless networks: joint flow routing and relay node assignment, in *IEEE INFOCOM*, pp. 1–9, 2010
22. B. Sirkeci-Mergen, A. Scaglione, On the power efficiency of cooperative broadcast in dense wireless networks. IEEE J. Sel. Areas Commun. **25**(2), 497 –507 (2007)
23. L. Tassiulas, A. Ephremides, Stability properties of constrained queueing systems and scheduling policies for maximum throughput in multihop radio networks. IEEE Trans. Automat. Control. **37**(12), 1936 –1948 (1992)

24. B. Wang, Z. Han, K. Liu, Distributed relay selection and power control for multiuser cooperative communication networks using buyer/seller game, in *IEEE INFOCOM*, pp. 544 –552, 2007

25. E. Yeh, R. Berry, Throughput optimal control of cooperative relay networks. IEEE Trans. Inf. Theory. **53**(10), 3827 –3833 (2007)

26. Y. Zhao, R. Adve, T. Lim, Improving amplify-and-forward relay networks: optimal power allocation versus selection. IEEE Trans. Wirel. Commun. **6**(8), 3114 –3123 (2007)

Chapter 7
Conclusion

Abstract In this chapter, we conclude this brief by summarizing our main contributions, and then present two possible future work directions: multihop cooperative D2D communication and online algorithm design.

7.1 Concluding Remarks

Data traffic in cellular networks has dramatically increased in recent years as the emergence of various new wireless applications, which imposes an immediate requirement for large network capacity. Although many efforts have been made to enhance the wireless channel capacity, they are far from solving the network capacity enhancement problem. Device-to-Device (D2D) communication is recently proposed as a promising technique to increase network capacity.

To increase D2D communication opportunities, we apply the cooperative communication to enhance the quality of D2D links in this brief. Specifically, we study the problem of maximizing the minimum transmission rate among multiple D2D pairs using cooperative communication in a cognitive radio cellular network. The relay assignment and channel allocation are jointly considered and network coding is exploited to improve the spectrum efficiency. Such max-min rate problems for cognitive and cooperative communications are proved to be NP-hard and the corresponding MINLP (Mixed-Integer Nonlinear Programming) formulations are developed.

Moreover, we apply the reformulation and linearization techniques to the original optimization problems with nonlinear and nonconvex constraints such that our proposed algorithms can produce high competitive solutions in a timely manner. After that, we study the energy efficiency of multiple D2D pairs with communication rate constraints using CC in multi-channel wireless networks. We prove it NP-hard and formulate it as a mixed-integer nonlinear programming (MINLP) problem, which is then transformed into a mixed-integer linear programming (MILP) problem using linearization and reformulation techniques. By exploiting several problem-specific characteristics, a time-efficient branch-and-bound algorithm is proposed to solve the MILP problem.

Finally, we study the relay selection problem with the objective of optimizing the throughput. For unicast, it is a P problem and an optimal relay selection algorithm is provided with a correctness proof. For broadcast, we show the challenge of

© The Author(s) 2014   77
P. Li, S. Guo, *Cooperative Device-to-Device Communication*
in Cognitive Radio Cellular Networks, SpringerBriefs in Computer Science,
DOI 10.1007/978-3-319-12595-4_7

relay selection by proving it NP-hard. A greedy heuristic algorithm is proposed to effectively choose a set of relay nodes that maximize the broadcast throughput.

7.2 Future Work

We can extend our work along several directions in future. First, we will extend our algorithms designed for single-hop cooperative D2D communication in this brief to deal with multihop cases. Due to the limited transmission power of mobile devices, they can only communicate with each other located within a small range. When multihop communication technique is adopted, each device can establish a data path with other nodes far away, leading to more D2D communication chances. On the other hand, multihop D2D communication involves more nodes for data forwarding, especially when cooperative relay is adopted. To achieve efficient multihop cooperative D2D communication, we need to deal with challenges in transmission scheduling, routing, and channel allocations.

Another direction of our future work is to design online algorithms to deal with network dynamics. In real cellular networks, mobile users will join or leave the network at any time, and the wireless channel quality also changes due to user mobility. Thus, it is important to design online algorithms to address these network dynamical events. The solutions in this brief can be used as theoretical bounds for online cases, and give us guideline in online algorithm design.